Occupational Safety and Health in the Emergency Services

Occupational Safety and Health in the Emergency Services

3rd Edition

James S. Angle

DELMAR
CENGAGE Learning·

Australia · Brazil · Japan · Korea · Mexico · Singapore · Spain · United Kingdom · United States

Occupational Safety and Health in the Emergency Services, Third Edition

James S. Angle

Vice President, Careers & Computing:
Dave Garza

Director of Learning Solutions:
Sandy Clark

Senior Acquisitions Editor:
Janet E. Maker

Managing Editor: Larry Main

Senior Product Manager:
Jennifer Starr

Editorial Assistant: Leah Costakis

Vice President, Marketing:
Jennifer Ann Baker

Marketing Director: Deborah Yarnell

Associate Marketing Manager:
Erica Glisson

Senior Production Director:
Wendy Troeger

Production Manager: Mark Bernard

Senior Content Project Manager:
Jennifer Hanley

Senior Art Director: Casey Kirchmayer

Library of Congress Control Number: 2012931467

ISBN-13: 978-1-4390-5750-6

ISBN-10: 1-4390-5750-8

Delmar
5 Maxwell Drive
Clifton Park, NY 12065-2919
USA

Cengage Learning is a leading provider of customized learning solutions with office locations around the globe, including Singapore, the United Kingdom, Australia, Mexico, Brazil, and Japan. Locate your local office at: **international.cengage.com/region**

Cengage Learning products are represented in Canada by Nelson Education, Ltd.

To learn more about Delmar, visit **www.cengage.com/delmar**

Purchase any of our products at your local college store or at our preferred online store **www.cengagebrain.com**

Notice to the Reader

Publisher does not warrant or guarantee any of the products described herein or perform any independent analysis in connection with any of the product information contained herein. Publisher does not assume, and expressly disclaims, any obligation to obtain and include information other than that provided to it by the manufacturer. The reader is expressly warned to consider and adopt all safety precautions that might be indicated by the activities described herein and to avoid all potential hazards. By following the instructions contained herein, the reader willingly assumes all risks in connection with such instructions. The publisher makes no representations or warranties of any kind, including but not limited to, the warranties of fitness for particular purpose or merchantability, nor are any such representations implied with respect to the material set forth herein, and the publisher takes no responsibility with respect to such material. The publisher shall not be liable for any special, consequential, or exemplary damages resulting, in whole or part, from the readers' use of, or reliance upon, this material.

Printed in the United States of America
1 2 3 4 5 6 7 16 15 14 13 12

Dedication

This book is dedicated to all those emergency responders who have been injured or killed in the line of duty and the valuable lessons they have taught us.

CONTENTS

Chapter 3   RISK MANAGEMENT / 55

Chapter 4   PREINCIDENT SAFETY / 71

Chapter 5   SAFETY AT THE FIRE EMERGENCY / 107

Chapter 6   SAFETY AT THE MEDICAL OR RESCUE EMERGENCY / 140

Chapter 7   SAFETY AT SPECIALIZED INCIDENTS / 157

Chapter 8   POSTINCIDENT SAFETY MANAGEMENT / 178

Chapter 9   PERSONNEL ROLES AND RESPONSIBILITIES / 194

Chapter 10   SAFETY PROGRAM DEVELOPMENT AND MANAGEMENT / 216

Chapter 11   SAFETY PROGRAM EVALUATION / 232

Chapter 12   INFORMATION MANAGEMENT / 240

Chapter 13   SPECIAL TOPICS IN EMERGENCY SERVICE OCCUPATIONAL SAFETY AND HEALTH / 259

APPENDICES

FOREWORD

It is an honor to write the forward for James Angle's *Occupational Safety and Health in the Emergency Services*. I consider Chief Angle to be not only a respected peer, but also a committed partner in the National Fallen Firefighters Foundation's efforts to reduce firefighter line-of-duty injuries and deaths.

The United States Congress established the National Fallen Firefighters Foundation in 1992 to honor our nation's fallen fire heroes and to provide their survivors with resources to help them rebuild their lives. When I was appointed Executive Director of the Foundation in July of 2001, no one could have predicted that within just a few months the events of September 11 would change both the fire service and the mission of the NFFF forever. The unprecedented fatalities suffered by the FDNY that day made many Americans conscious of the tremendous risks that emergency responders face every day. As we made every effort to reassure the thousands of survivors that their loved ones would never be forgotten, the Foundation was impacted in ways that we had never previously imagined.

That experience also reminded us that each year, approximately 100 firefighters die in the line of duty. That translates to more than 3,200 firefighters who have been honored on the National Fallen Firefighters Memorial since it was established in 1981, leaving behind thousands of bereaved family members and coworkers to rebuild their lives. The lessons that we learned in New York after 9/11 continue to influence almost everything that we do at the Foundation, and have helped to vastly improve the services that we provide to these survivors.

In my previous career as a firefighter, officer, and fire chief, I had become painfully aware that some of the men and women who were honored by the Foundation had died in ways that were preventable. As I spoke with leaders within the fire service, I heard from others who had witnessed similar events. Many were beginning to ask some very tough questions, including *why do we accept this as a culture?* and *what can we do to prevent it?*

As these conversations continued, the United States Fire Administration responded by establishing a goal of reducing firefighter line-of-duty deaths by 50% within a decade. In March of 2004, the NFFF supported this effort by coordinating the first Firefighter Life Safety Summit in Tampa, Florida. Representatives of every identifiable segment of the fire service attended the Summit, including those of all major national fire service organizations, as well as occupational health and safety experts and survivors of fallen firefighters. For two days, participants deliberated over cultural, philosophical, technical, and procedural problems existing within the fire service. Together we identified the most important causal factors influencing line-of-duty deaths, which were eventually clustered into six domains. During the Summit, more than one hundred recommendations were discussed that were then narrowed into a set of key initiatives that would define the mission of this effort going forward. From there, the *Firefighter Life Safety Initiatives* were agreed upon, and the Everyone Goes Home® Program was born.

Much of the information contained within this textbook focuses on ways that fire departments can utilize the *Firefighter Life Safety Initiatives*. They serve as a starting point for making occupational health and safety a personal priority for every employee or member, and an organizational priority for all sizes of fire departments, from the very smallest to the largest Metro agency. My hope is that you will use the Initiatives outlined in this book as guiding principles by which to examine and improve the ways that you and your coworkers perform your job, focusing on health and safety. You will also learn how to address any existing deficiencies within your department's organizational structure and to identify operational procedures that may put you and your fellow firefighters at risk. From there, you will begin to understand how to integrate firefighter health and safety into the very culture and ethos of your department.

You will also learn that Occupational Health and Safety is a comparatively new field for the fire service industry. Our late adoption of this science has enabled us to take advantage of work done in other high-risk industries, such as the military and emergency response organizations in other countries, and to adapt some of their ideas and methodologies for use in the fire service. This knowledge transfer has resulted in significant advancements in firefighter health and safety in a relatively short time period, including the integration of risk management into fire department operations, and many of the behavioral health support programs proffered by the National Fallen Firefighters Foundation.

In this book you will see that Occupational Health and Safety is a dynamic and changing science. Throughout your career, it will be your responsibility to stay current in relation to OSHA regulations, NFPA standards, and the most up-to-date information pertaining to best practices. Prioritizing firefighter safety is about having the mental tools—education, training, and experience—that enable you to accurately assess a situation, make an appropriate risk management decision, and minimize the likelihood that you or a member of your team will suffer a preventable injury or even death. Far too often, after a line-of-duty death, we hear from the company and chief officers involved that "we never thought it could happen to us." Within this text, you will review case studies that clearly demonstrate that injuries and fatalities must always be seen as a potential outcome of virtually any fire department activity. Lack of preparation is unconscionable, because if you are not adequately prepared, a line-of-duty injury or fatality can—and very likely will—happen to you or a fellow firefighter.

As you develop and utilize your knowledge of Occupational Safety and Health, you will be supporting a wave of change, and helping to create a "Culture of Safety" that is becoming the new standard in fire departments across the United States. The concept of fostering a safety culture is not new, but instead emphasizes what has always been a core principal of the more global "Culture of Firefighting." It reinforces that for firefighters, the preservation of life, whether it is the lives of the public we serve or those of our fellow firefighters, is our most important mission.

As the next generation of firefighters and fire officers, you and your peers are charged with a very important task. The nation's fire service is looking for you to continue the work begun by Chief Angle, myself, and the Tampa Summit attendees, integrating the *Firefighter Life Safety Initiatives* into the mission, culture, policies, and procedures of every fire service organization. Hopefully yours will be the first generation to experience a future in which every firefighter does indeed go home to their family at the end of every shift.

Chief Ronald J. Siarnicki
Executive Director
National Fallen Firefighters Foundation

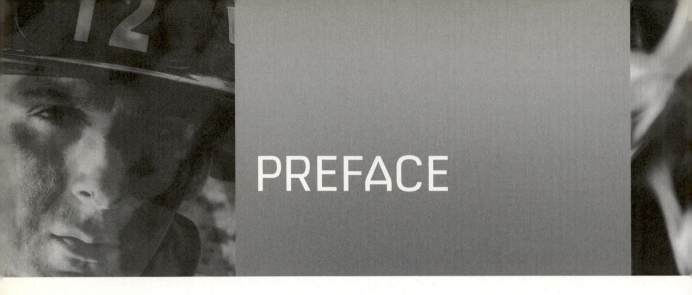

PREFACE

Each year, a number of emergency responders are killed or injured in the line of duty. It is my hope that this book will have a significant, positive impact on these statistics. Since the publication of the previous editions, much has changed for emergency service responders. Emphasis continues on preparedness for attacks using weapons of mass destruction. Technology has been improved in terms of tools and equipment used by responders. Additional standards and regulations have been promulgated in an effort to further the safety and health of responders. Other standards and regulations have been updated. While the emphasis on occupational safety and health in fire and emergency medical services is a fairly new discipline, much can be learned by the concepts that have been in place in industry for many years. The increased emphasis, and the increase in accountability for safe practices over the last 20 years creates the need for education and training in safety and health practices that focuses on both fire and emergency medical services operations.

During the development of the second edition of this text, a historic event took place. In Tampa, Florida, during March 2004, the National Fallen Firefighters Foundation hosted a 2-day symposium on preventing firefighter line-of-duty deaths. This symposium brought together 200 fire service leaders from across the country to discuss the issue of line-of-duty-deaths. The group looked at the line-of-duty-death problem from six perspectives: reducing fires, training, structure fires, vehicle safety, wildland fires, and wellness and fitness. After a day of working in subgroups, findings were presented to the full group. The summary of this work resulted in the 16 Firefighter Life Safety Initiatives. This author had the honor—and, considering the attendees, it truly was an honor—to attend and participate in this event. Further information from the symposium can be found in Appendix C. The third edition of this text includes the relationship of these 16 Firefighter Life Safety Initiatives to the information being presented in each of the chapters.

ABOUT THIS BOOK

Many colleges offer credit courses in this area as part of degree programs or as stand-alone courses. Additionally, this book is intended for such a course in emergency service occupational safety and health. The Fire and Emergency Services Higher Education group recommends an Emergency Safety and Health course as part of the associate degree curriculum. Further, fire and EMS departments can use this book as a reference and for in-house instruction in safety and health programs. Students should also be encouraged to read emergency services trade journals and websites because they frequently have articles focusing on safety and health issues.

This third edition continues with the comprehensive approach to emergency service occupational safety and health. The purpose of this book is to provide instructors, students, safety officers, fire and EMS department managers, and others in the emergency services field a one-stop resource for total safety program management.

The text introduces the reader to occupational safety and health from a historical standpoint, including a historical look at industrial safety and health. It then guides the reader through a process that will allow them to put the information to work within their organizations. The outcome, it is hoped, will be a safer work environment that will reduce costs associated with responder injuries and deaths and increase productivity.

HOW TO USE THIS BOOK

This text is designed to be used for a semester or quarter-hour college-level course on occupational safety and health within the fire or EMS discipline. The book was also designed to be used by fire and EMS agencies—whether fully paid, combination, or volunteer—as an in-house reference, a manual for an organizational training program, or as a promotional testing reference. The text is designed to give the reader a broad overview of issues to consider for various types of incidents common to emergency response.

CHAPTER 1 provides a brief history of occupational safety and health for emergency services, with a brief examination of safety and health in industry over the years. It gives the reader an understanding of the importance of and the rationale for the increasing emphasis on occupational safety and health over the past 20 years.

CHAPTER 2 defines and reviews standards and regulations that have an impact on safety programs.

CHAPTER 3 presents risk analysis and management to allow the reader to form an understanding of this important concept before specific situations are presented.

CHAPTERS 4–8 focus on specific safety issues including preincident, fire, medical/rescue, and specialized incidents, as well as postincident implications.

CHAPTER 9 focuses on issues relating to personnel roles and responsibilities. Personnel issues are often the most difficult to deal with in terms of safety programs.

CHAPTERS 10–12 deal with program management, including development and ongoing management, evaluation of effectiveness, and information management.

CHAPTER 13 includes a look at current legal, ethical, and financial issues as well as future trends.

FEATURES OF THIS BOOK

The following features are included in this book to ensure that the contents of the chapters are reader friendly and to promote understanding about the importance of safety in the emergency services:

■ A CASE STUDY opens each chapter and is designed to stimulate discussion. The stories are drawn from actual events and stress the importance of learning how to prevent accidents and fatalities and to advance optimum health and safety in the emergency services.

■ LEARNING OBJECTIVES are clearly outlined at the beginning of each chapter to highlight what the student should expect to learn on completion of the chapter.

■ KEY TERMS AND NOTES are integrated into each chapter to highlight important points and introduce new concepts.

■ REVIEW QUESTIONS based on the learning objectives presented at the beginning of the chapter are included at the end of each chapter, allowing students to assess the knowledge they have gained.

■ ACTIVITIES at the end of each chapter are designed to allow students to put into practice the knowledge they have gained through study of the concepts presented.

NEW TO THIS EDITION

This book was thoroughly updated with the latest information, practices, and trends, including the following:

■ ALL-NEW DESIGN AND NEW PHOTOS engage the student and focus on current practices in the emergency services.

■ NEW CASE STUDIES reflect current events and timely issues for emergency responders.

■ 16 FIREFIGHTER LIFE SAFETY INITIATIVES are emphasized throughout the chapters, including correlations that illustrate where each chapter relates to specific Firefighter Life Safety Initiatives.

- **NEW INFORMATION** related to air management, medical issues, current technologies, highway scene safety, and other crucial topics prepare students for safe emergency response.
- **CURRENT NFPA STANDARDS AND OSHA REGULATIONS**, including *NFPA Standards 1500* and *1710*, highlight the latest requirements for responders.
- **CURRENT STATISTICS** reflect trends in the emergency services and keep students informed of the latest issues affecting occupational safety and health.

SUPPLEMENT TO THIS BOOK

An *Instructor Resources CD* is available that contains many features to aid the instructor in classroom preparation:

- **LESSON PLANS** detail the topics in each chapter and correlate to the accompanying PowerPoint presentations, to ensure that instructors highlight important points of discussion. (Editable)
- **ANSWERS TO QUESTIONS** offer a quick reference to validate learning.
- **POWERPOINT PRESENTATIONS** align with the Lesson Plans and provide graphics and photos to visually enhance classroom presentations. (Editable)
- **EXAMVIEW® TEST BANKS** are available for each chapter, allowing instructors to revise questions or create new tests based on the existing questions. (Editable)
- **ADDITIONAL CASE STUDIES** are available to instructors as handouts for further classroom discussion on important topics.
- **IMAGE GALLERY** offers additional photos and graphics from the book to enable instructors to create their own presentations, or add to existing presentations.
- **FESHE AND NFPA CORRELATIONS** illustrate where content in the book aligns with the FESHE course outcomes for Occupational Safety & Health for Emergency Services and applicable NFPA standards.

ABOUT THE AUTHOR

James S. Angle is the fire chief of the Palm Harbor Fire Rescue Department in Pinellas County, Florida. Internationally accredited by the Commission on Fire Accreditation International, the department protects 60,000 people in a 20-square-mile area operating from four fire stations, and provides a full range of services including fire prevention, public education, advanced life support first response, rescue, hazardous materials response, and fire suppression. He is a 37-year emergency service veteran, having begun his career in the Monroeville Fire Department, Company 4, in suburban Pittsburgh, Pennsylvania. He also served with Pittsburgh Emergency Medical Services for 5 years as a paramedic crew chief. His background includes employment with four fire rescue agencies (both small and large) in the state of Florida including Hallandale Beach Fire Rescue Department, Palm Beach County Fire Rescue, South Trail Fire Rescue, and Palm Harbor Fire Rescue. He has worked through the ranks from firefighter/paramedic to his current position as fire chief, including 8 years as a training and safety officer.

His education includes an associate degree in Fire Science Administration from Broward Community College, a bachelor's degree in Fire Science and Safety Engineering from the University of Cincinnati, and a master's degree in Business from Nova University. He also holds Instructor III certification from the Florida Bureau of Fire Standards and Training, is certified as a paramedic, and is a graduate of the Executive Fire Officer Program of the National Fire Academy. He also holds Chief Fire Officer Designation from the Commission on Chief Fire Officer Designation.

As an instructor, he teaches various fire science classes for St. Petersburg College, including a three-credit occupational safety and health class. He has delivered seminars to national audiences at three Fire Department Instructors Conferences as well as conferences held by the Florida Fire Chiefs Association.

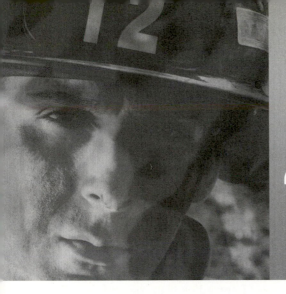

ACKNOWLEDGMENTS

This section could be a chapter in and of itself. I would be remiss if I did not begin with thanks to my wife Joann, and my two sons, Anthony and Austin. When undertaking a project such as this, the support of the family is one of the most important needs. My lovely wife and great kids have always been there to lend a helping hand or provide some words of encouragement. For this, I once again give them my sincerest thanks. When the first and second editions were written, the kids were kids. As this third edition is written the kids are young men. In fact the oldest has begun his fire service career with the Marion County, Florida, Fire Rescue Department. The youngest is well on his way and currently serves as chief in his Fire Explorer post. My wife JoAnn has always been the glue that holds us all together. THANKS.

© Cengage Learning 2013

I would like to thank my father, who passed away during the writing of the first edition of this book, as he is the one who took me to my first fire, which at that moment made me realize that I never wanted to do anything other than work in this field.

I owe a special thanks to EMS/Safety Chief Bob Markford of Palm Harbor Fire Rescue who, among his varied other duties at our emergencies, sometimes finds a minute to get some outstanding photos, many of which are used in this text.

I could name many individuals who had an impact on me throughout my career, but the list would be longer than the book. So many have helped me to make this my career, and not just a job. I would like to recognize the Board of Fire Commissioners and the members of the Palm Harbor Fire Rescue Department, who have provided encouragement and support throughout this endeavor. I also thank the management and members of the South Trail Fire Department, where for 8 years I served as training and safety officer, and where I realized that through a good process, management support, and member support, an organization can reduce injuries. As for the other departments in which I had the pleasure of serving, be assured that I learned something at each of them that I use today, and it can be found somewhere in this book.

A special thanks to Mark Huth of Delmar Cengage Learning, who took a chance on the first edition and allowed me to make a dream a reality. Also a special thanks to Jennifer Starr as the Developmental Editor for all of her help, encouragement, and weekly phone calls to make sure that I was on track—even though I was often late—along with appreciation to all of the production staff at Delmar Cengage Learning.

In addition, I would thank the reviewers whose time and careful review have ultimately made this a better book:

Timothy Baker
Fire Science Program Instructor
Lansing Community College
Lansing, MI

Dr. Gareth C. Burton
McDowell Owens Engineering
Kingwood, TX

Dave Casey, MPA, EFO, Bureau Chief
Bureau of Fire Standards, Training & Safety
Florida Fire Division of State Fire Marshal
Ocala, FL

Richard A. Dienst
Fire Science Program Manager
Tidewater Community College
Virginia Beach, VA

Paul Grant, MA Ed
Red Rocks Community College
Lakewood, CO

Dennis Hopkins
Caldwell Community College
Hudson, NC

Gail Hughes, Assistant Professor
Chattanooga State Tech CC
Chattanooga, TN

Tom C. Jenkins
Fire Chief
Rogers Arkansas Fire Department
Rogers, AR

Dr. Bill Lowe, E.M.T. – P
Clayton County Fire Department
Fayetteville, GA
Richard C. Maddox
Health & Safety Officer
Sayville Fire Department
Sayville, NY

Michael Petroff
Chairman, Board of Directors
Fire Department Safety Officers Association
Ashland, MA

William Shouldis
Director of Training
Philadelphia Fire Department
Philadelphia, PA

Deputy Chief James P. Smith
Philadelphia Fire Department
Philadelphia, PA

Tom Wutz
Chief Fire Services (Retired)
New York State Office of Fire Prevention and Control
Albany, NY

FIRE AND EMERGENCY SERVICES HIGHER EDUCATION (FESHE)

In June 2001, The U.S. Fire Administration hosted the third annual Fire and Emergency Services Higher Education Conference, at the National Fire Academy campus, in Emmitsburg, Maryland. Attendees from state and local fire service training agencies, as well as colleges and universities with fire related degree programs attended the conference and participated in work groups. Among the significant outcomes of the working groups was the development of standard titles, outcomes, and descriptions for six core associate-level courses for the model fire science curriculum that had been developed by the group the previous year.

The six core courses are Fundamentals of Fire Protection, Fire Protection Systems, Fire Behavior and Combustion, Fire Protection Hydraulics and Water Supply, Building Construction for Fire Protection, and Fire Prevention.* The committee also developed similar outlines for other recommended courses offered in fire science programs. These courses included: Fire Administration I, Occupational Health and Safety, Legal Aspects of the Emergency Services, Hazardous Materials Chemistry, Strategy and Tactics, Fire Investigation I, and Fire Investigation II.

FESHE CONTENT AREA COMPARISON

The following table provides a comparison of the 12 FESHE content areas with this text.

2001 Fire and Emergency Services Higher Education Conference Final Report, (Emmitsburg, Maryland: U.S. Fire Administration) 2001, PAGE 12.

COURSE CORRELATION GRID

OCCUPATIONAL HEALTH AND SAFETY

Course Description This course introduces the basic concepts of occupational health and safety as it relates to emergency service organizations. Topics include risk and hazard evaluation and control procedures for emergency service organizations.

Outcomes		Occupational Safety and Health in the Emergency Services 3rd Edition Chapter Reference
1	Describe the history of occupational health and safety.	1
2	Identify occupational health and safety programs for industry and emergency services today.	1
3	Compare the difference between standards and regulations.	2
4	List and describe the components of risk identification, risk evaluation, and incident management.	3
5	Describe the relevance for safety in the work place including the importance of PPE.	5, 6, 7
6	Apply the knowledge of an effective safety plan to preincident planning, response, and training activities.	4
7	Explain the components of an accountability system in emergency service operations.	5
8	Discuss the need for and the process used for postincident analysis.	8
9	Describe the components and value of critical incident management programs.	8
10	Describe the responsibilities of individual responders, supervisors, safety officers, and incident commanders, safety program managers, safety committees, and fire department managers as they relate to health and safety programs.	9
11	Describe the components of a wellness/fitness plan.	4
12	Identify and analyze the major causes involved in line-of-duty firefighter deaths related to health, wellness, fitness, and vehicle operations.	1,4

NFPA 1500 STANDARD CORRELATION GRID

NFPA 1500 Chapter	Occupational Safety and Health in the Emergency Services 3rd Edition Chapter Reference
1 Administration	10
2 Referenced Publications	2
3 Definitions	
4 Fire Department Administration	3, 11, 12
5 Training and Education	4, 10
6 Fire Apparatus, Equipment, and Driver/Operators	4
7 Protective Clothing and Protective Equipment	5
8 Emergency Operations	5
9 Facility Safety	4
10 Medical and Physical Requirements	4
11 Member Assistance and Wellness Programs	4
12 Critical Incident Stress Program	8

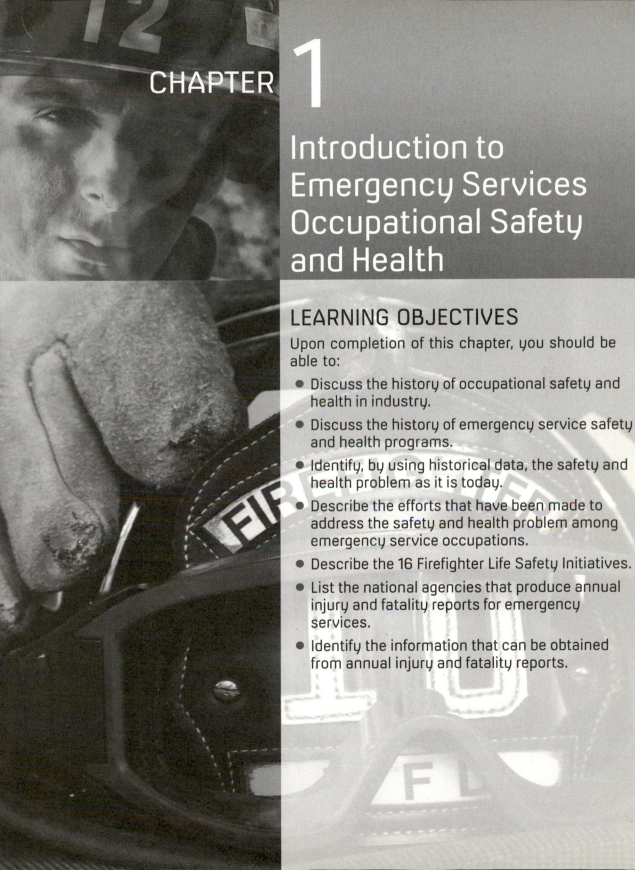

CHAPTER **1**

Introduction to Emergency Services Occupational Safety and Health

LEARNING OBJECTIVES

Upon completion of this chapter, you should be able to:

- Discuss the history of occupational safety and health in industry.
- Discuss the history of emergency service safety and health programs.
- Identify, by using historical data, the safety and health problem as it is today.
- Describe the efforts that have been made to address the safety and health problem among emergency service occupations.
- Describe the 16 Firefighter Life Safety Initiatives.
- List the national agencies that produce annual injury and fatality reports for emergency services.
- Identify the information that can be obtained from annual injury and fatality reports.

On March 30, 2010, a firefighter/paramedic (FF # 1) died and a firefighter/paramedic (FF # 2) was injured when caught in an apparent flashover while operating a hoseline within a residence. Units arrived on scene to find heavy fire conditions at the rear of a house and moderate smoke conditions within the uninvolved areas of the house. A search and rescue crew had made entry into the house to search for a civilian who was reportedly entrapped at the rear of the house. FF # 1, FF # 2, and a third fire fighter made entry into the home with a charged 2½-inch hoseline. Thick, black, rolling smoke banked down to knee level after the hoseline was advanced 12 feet into the kitchen area. While ventilation activities were occurring, the search and rescue crew observed fire rolling across the ceiling within the smoke. They immediately yelled to the hoseline crew to "get out." The search and rescue crew were able to exit the structure safely, then returned to rescue FF # 2 first and then FF # 1. FF # 1 was found wrapped in the 2½-inch hoseline, which had ruptured, and without his facepiece on. He was quickly brought out of the structure, received medical care on scene, and was transported to a local hospital where he was pronounced dead.

Contributing Factors to the Firefighter's Injury and Death:

- Well-involved fire with entrapped civilian upon arrival
- Incomplete 360-degree situational size-up
- Inadequate risk-versus-gain analysis
- Ineffective fire control tactics
- Failure to recognize, understand, and react to deteriorating conditions
- Uncoordinated ventilation and its effect on fire behavior
- Removal of self-contained breathing apparatus (SCBA) facepiece
- Inadequate command, control, and accountability
- Insufficient staffing

This fire occurred in a 950 square foot single-family residence. It illustrates that responder deaths and injuries can occur anywhere, anytime, and under any circumstances. Having a comprehensive occupational safety and health program is essential in all emergency service agencies.

Source: Information for this case study was taken in part from CDC/NIOSH Fire Fighter Fatality Investigation Report F2010-10. http://www.cdc.gov/niosh/fire/reports/face201010.html

INTRODUCTION

Firefighting in the United States is sometimes described as one of the nation's most dangerous occupations because of the high rate of acute and chronic injuries, illness, and deaths. Firefighters are exposed to a wide range of hazards including chemicals, carcinogens, extreme weather, building collapses, smoke, heat, physical and emotional stress, and traveling under less-than-ideal roadway conditions. It is no wonder that the occupation suffers a high rate of injury and death.

Emergency medical service responders face many of the same hazards as firefighters. In fact, emergency medical responders face additional exposure, including concerns such as infectious diseases. Most firefighters are also responsible for emergency medical response. In fact, the fire service is the nation's largest provider of prehospital care.[1] Because of this close relationship between the firefighters and the responders to emergency medical incidents, a complete occupational safety and health program must focus on both fires and medical emergencies. Therefore, an examination of the problems and the methods to improve a historically poor record is conducted from the emergency services standpoint.

HISTORY OF OCCUPATIONAL SAFETY AND HEALTH IN INDUSTRY

Occupational safety and health in the workplace is certainly not a new concept—in fact, the history can be traced back thousands of years. Many important people throughout history have contributed to the study of **occupational diseases** and **occupational injuries**. Hippocrates (460–377 BC) described symptoms of lead poisoning among miners. Paracelsus (1493–1541) wrote a treatise on occupational diseases. He described lung diseases among miners and attributed the cause to vapors from metals. He also is known for his work in toxicology because of his observations of dose and response. Agricolos (1494–1555) wrote about the need to provide ventilation for miners. Bernardino Ramazzini (1633–1714) was known as the patron saint of industrial medicine. He had advised physicians to learn about occupational disease by studying the work environment, and to always ask their patients their trade.

Over time, as these and many others were making strides in industrial occupational safety and health, many events also occurred that highlighted the need for change in workplace safety. Often people had to fight for better and safer working conditions. These fights involved strikes, lockouts, fighting in the streets, and in some cases prison sentences. One famous event that most responders should be familiar with is the 1911 fire at the Triangle Shirtwaist Company in New York City. This fire claimed the

occupational disease An abnormal condition, other than an occupational injury, caused by an exposure to environmental factors associated with employment.

occupational injury An injury that results from exposure to a single incident in the work environment.

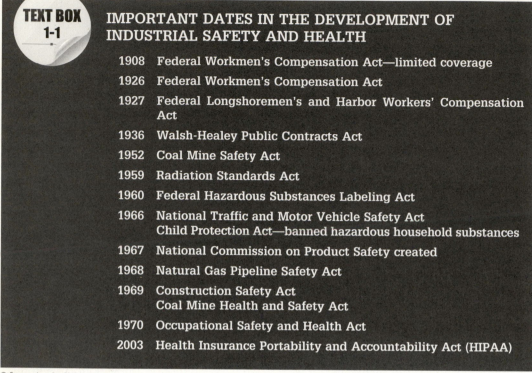

TEXT BOX 1-1

IMPORTANT DATES IN THE DEVELOPMENT OF INDUSTRIAL SAFETY AND HEALTH

1908	Federal Workmen's Compensation Act—limited coverage
1926	Federal Workmen's Compensation Act
1927	Federal Longshoremen's and Harbor Workers' Compensation Act
1936	Walsh-Healey Public Contracts Act
1952	Coal Mine Safety Act
1959	Radiation Standards Act
1960	Federal Hazardous Substances Labeling Act
1966	National Traffic and Motor Vehicle Safety Act
	Child Protection Act—banned hazardous household substances
1967	National Commission on Product Safety created
1968	Natural Gas Pipeline Safety Act
1969	Construction Safety Act
	Coal Mine Health and Safety Act
1970	Occupational Safety and Health Act
2003	Health Insurance Portability and Accountability Act (HIPAA)

© Cengage Learning 2013

lives of 146 workers and called attention to the use of cramped and overcrowded working conditions and lack of exits. In 1914, studies in New York and Ohio highlighted tuberculosis and unsanitary conditions, leading to the abolishment of sweatshops. Studies in 1923 of "dusty trades" led to the development of industrial hygiene sampling equipment.

These notable events and studies led to the development of laws, regulations, and standards to better protect workers. **Text Box 1-1** shows some significant milestones in the development of federal occupational safety and health legislation.

As a result of this history and the need for safe working conditions, a multitude of programs exist today in industry. These programs deal with air quality, hearing protection, and ergonomics, to name a very few. Many of these programs have been developed because they are required by regulations. Industrial facilities have safety and health divisions; colleges and universities grant degrees at every level for industrial hygiene and occupational safety and health. Regulations and standards set forth the minimum requirements for equipment and procedures. Facilities and equipment are inspected for safe operation, and personnel have access to medical surveillance in many occupations in order to find diseases early.

Emphasis on occupational safety and health in the emergency services is a new concept when considering the history of industrial safety and health described previously. However, in the past 30 years it has been recognized that these age-old concepts must also be applied to firefighters and EMS responders.

HISTORY OF EMERGENCY SERVICES SAFETY AND HEALTH

For many years, the high injury and death rate for firefighters was accepted as part of the occupation (**Figure 1-1**). As firefighters moved into the role of EMS responders, the injury potential increased, not only because of additional exposures, but also because the increase in incidents led to a higher potential for responding and returning accidents.

However, fire and EMS occupational safety and health have matured over the past 30 years (see **Text Box 1-2**). As in industry, the development of regulations and standards often came after a tragic event or through the study of occupational injuries and occupational diseases.

Many of the early fire service texts used in the education and training of firefighters and EMS responders make little or no reference to the injury and death problem from a standpoint of improvement. There are references to the dangers of the profession; however, little is provided for prevention strategies.

One of the first publicized documents making reference to firefighter safety was *America Burning*.[2] In this 1973 report to the president of the United States, the National Commission on Fire Prevention and Control reported on the nation's fire problem (see **Figure 1-2**). One section of the report focused on firefighter safety from the perspective of staffing, education, and equipment. The 1980s saw a significant increased interest in the safety problem facing the emergency responders, and this trend has continued into

FIGURE 1-1
Modern fire apparatus design has a focus on responder safety.

TEXT BOX 1-2

IMPORTANT DATES IN THE DEVELOPMENT OF EMERGENCY SERVICE OCCUPATIONAL SAFETY AND HEALTH

1960   International Association of Firefighters begins to publish a Firefighter Injury and Death Report

1973   *America Burning* published

1974   NFPA begins to publish an Annual Firefighter Injury report and an Annual Firefighter Death Report

1986   *NFPA 1403, Standard on Live Fire Training Evolutions* adopted by NFPA

1987   *NFPA 1500, Standard on Fire Department Occupational Safety and Health Programs* adopted by NFPA

1992   The United States Congress creates the National Fallen Firefighters Foundation

1996   OSHA ruling on 2 In/2 Out procedures referencing OSHA 1910.134

1997   Joint IAFC/IAFF Wellness/Fitness program developed

1998   NIOSH Fire Fighter Fatality Investigation and Prevention Program began

1998   OSHA Respiratory Protection Standard applied to firefighters

2001   *NFPA 1710, Standard for the Organization and Deployment of Fire Suppression Operations, Emergency Medical Operations, and Special Operations to the Public by Career Fire Departments* adopted by NFPA

2004   Firefighter Life Safety Summit—Development of the 16 Firefighter Life Safety Initiatives

© Cengage Learning 2013

The Commission on Fire Prevention and Control has made a good beginning, but it cannot do our work for us. Only people can prevent fires. We must become constantly alert to the threat of fires to ourselves, our children, and our homes. Fire is almost always the result of human carelessness. Each one of us must become aware—not for a single time, but for all the year—of what he or she can do to prevent fires.

—*President Richard M. Nixon*
September 7, 1972

© Cengage Learning 2013

FIGURE 1-2
President Nixon's preface to *America Burning.*

the 21st century. One of the most controversial—yet positive—steps in the safety and health area was the publication in 1987 of **NFPA 1500**, *Standard on Fire Department Occupational Safety and Health Program*. This standard and the series of other safety standards that followed set the safety and health movement in motion. The standard was controversial because of potential changes in operations and costs, but it caused managers, unions, volunteer fire chiefs, and others to note that the statistics for death and injury were unacceptable, and something had to be done.

NFPA 1403, Standard on Live Fire Training Evolutions was adopted in 1986, following a loss of life in fire training in 1982. This standard is the minimum standard for live fire-training exercises, whether they are held in acquired structures or at training centers. In 1996, the Occupational Safety and Health Administration made the ruling on the two-in/two-out procedures (see Appendix B for additional information), which was followed in 1998 by an updated respiratory protection standard. In 2001, *NFPA 1710* was approved by the National Fire Protection Association (NFPA). While primarily a deployment standard, many of the requirements dealt with issues relating to firefighter/EMS responder safety and health. These are described in more detail in the following chapters.

The past ten years also saw the nation's responders becoming more prepared for the threat of terrorism and attacks by weapons of mass destruction. Following the attacks on September 11, 2001, fire and EMS providers had to step up their preparedness levels. These new threats to the safety and health of the responders caused shifts in thinking and highlighted the need to acquire new equipment and to provide a new, higher level of training for this kind of event. The emergency services were called on later in 2001 for anthrax events, furthering the concern for the safety and health of responders from yet a different hazard.

Emergency services organizations are often reluctant to make and accept change. However, procedures, techniques, and often equipment do change. Often these changes are engineered without considering the human element of the change. For example, a department can implement the best all-hazard incident management system with scene accountability, but if individual companies or crews freelance within the system, the program will fail. All too often, the attitude of the response personnel mimics that of our nation's population when it comes to fires and injuries—*it won't happen to me*. However, it *does* happen to a firefighter almost 100,000 times a year for injuries and almost 100 times a year for deaths.

Since the publication of *NFPA 1500*, emergency services organizations have seen changes. It is now common to find organization charts that include a health and safety officer, new textbooks and new editions with sections on health and safety, safety and

---

NFPA 1500 The National Fire Protection Association's consensus standard on fire department occupational safety and health.

health committees formed with representation from all levels of the organization, and standard operational procedures intended to provide a safe working environment. Incident management systems have been established, the use of personal protective equipment is required, and safer equipment has been developed. All of this has been accomplished with the goal of improved safety and health for the responder. Many emergency service organizations offer yearly medical examinations in addition to the examination at time of hire. These regular physicals allow the formation of a baseline for an annual comparison of medical history over the length of a member's career. This trend will and should continue.

NOTE Another significant improvement since the movement began is the recognition that the local injury problem should be evaluated.

National statistics have been gathered about occupational injuries, but often this information is not compared to local statistics until a problem or tragic event occurs.

IDENTIFICATION OF THE SAFETY PROBLEM

A number of resources are available to help a Safety and Health officer develop a program. Clearly, local program design and development must focus on local problems. It would not make sense for Florida departments to invest considerable money to prevent injuries associated with exposure to cold. Likewise, a department providing EMS transport that experiences a high number of back injuries may want to make a substantial investment in a back injury prevention program. Local statistics can be gathered, and the data can be used to determine the local safety and health problem. Techniques for gathering this information and the process for the analysis are provided in Chapter 12. However, local data should be compared to that of the larger population from the state and national figures. The following organizations publish annual safety and health data that can be useful in this endeavor.

National Fire Protection Association

In addition to publishing standards, the NFPA publishes annual reports on both occupational injuries and deaths in the fire service in its *Fire Journal* magazine. The NFPA has been compiling these reports since 1974. The NFPA death survey is a report on all firefighter deaths and includes analysis of the fatalities in terms of type of duty, cause of death, age group comparisons, and population-served comparisons. The NFPA injury survey is not an actual survey of all departments; rather, it is a sample used to project the national firefighter injury experience. Although a prediction, the survey inspires a high level of confidence and is representative of all sizes and types of departments. These reports are very comprehensive and useful for the safety program manager to compare national to local experience.

United States Fire Administration

The **United States Fire Administration (USFA)** oversees the **National Fire Incident Reporting System (NFIRS)**. Part of the NFIRS report's purpose is to collect data for firefighter casualties. This data can be obtained through the USFA and can be used to compare national data to local data. However, there are some problems with this system. The NFIRS is a voluntary system in which departments may or may not participate, which creates the problem of not knowing what injuries or deaths may go unreported because a local jurisdiction did not participate in the program. Further, as the name implies, the report is used for incidents; therefore, injuries that occur to on-duty personnel outside of incidents are not captured.

The NFIRS data is formulated into reports in various formats and is available by mail from the USFA publication center. Much emergency service safety and health data is available, including case analyses of incidents in which fatalities or injuries occurred as well as protective clothing field test results. These results are also available from the USFA publication center (see **Figure 1-3**). The USFA also has an intensive

FIGURE 1-3
Overview of the number of on-duty deaths and injuries that occurred from 2001–2010.

Firefighter Casualties 2001–2010

The following table gives an overview of the number of on-duty deaths and injuries that occurred from 2001 to 2010.

Year	Deaths[1]	Fireground Injuries[2]	Total Injuries[2]
2001	105[3]	41,395	82,250
2002	101	37,860	80,800
2003	113	38,045	78,750
2004	119[4]	36,880	75,840
2005	115	41,950	80,100
2006	107	44,210	83,400
2007	118	38,340	80,100
2008	118	36,595	79,700
2009	90	32,205	78,150
2010	87	32,675	71,875

[1] This figure reflects the number of deaths as published in USFA's annual report on firefighter fatalities. All totals are provisional and subject to change as further information about individual fatality incidents is presented to USFA.
[2] This figure reflects the number of injuries as published in NFPA's annual report on firefighter injuries.
[3] In 2001, an additional 341 FDNY firefighters, three fire safety directors, two FDNY paramedics, and one volunteer from Jericho (NY) Fire Department died in the line of duty at the World Trade Center on September 11.
[4] The Hometown Heroes Survivors Benefit Act of 2003 has resulted in an approximate 10% increase to the total number of firefighter fatalities counted for the annual USFA report on Firefighter Fatalities in the United States beginning with CY2004. For more information, please see Reports and Statistics.

Source: United States Fire Administration

website that provides a means to order publications, but many are downloadable and can be reviewed immediately.

United States Fire Administration Department under the Federal Emergency Management Agency that directs and produces fire programs, research, and education.

National Fire Incident Reporting System (NFIRS) The uniform fire incident reporting system for the United States; the data from this report is analyzed by the United States Fire Administration.

International Association of Fire Fighters

Since 1960, the **International Association of Fire Fighters (IAFF)** has produced an annual firefighter injury and death study. This research reports on a number of issues relating to safety and health, similar to that of the NFPA. However, the report is more in-depth and provides information on lost-time injuries and exposure to infectious diseases. This report is another resource for the safety and health program manager, but the data is only gathered from paid fire departments that have IAFF affiliation, which limits the study to these departments. In 1997, the IAFF, along with the **International Association of Fire Chiefs (IAFC)**, presented the Fire Service Joint Labor/ Management Wellness-Fitness initiative. This initiative contains a component of data reporting to the IAFF on employee wellness, fitness, and injury issues. This report will be a great resource to the safety manager when sufficient data is received, but it will only contain data from departments that choose to participate.

International Association of Fire Fighters (IAFF) Labor organization that represents the majority of organized firefighters in the United States and Canada.

International Association of Fire Chiefs (IAFC) Organization of fire chiefs from the United States and Canada.

Occupational Safety and Health Administration

The **Occupational Safety and Health Administration (OSHA)** is an agency within the Department of Labor. OSHA regulations are applicable to many public fire departments and to all private fire departments (see Chapter 2 for additional information on OSHA regulations applicable to public agencies in your state). OSHA mandates certain reporting requirements for occupation-related injuries and deaths. This requirement for record keeping and reporting allows OSHA to compile useful statistics and to study causes of occupational injuries in order to develop prevention strategies and countermeasures. Because not every state requires that public fire departments comply with OSHA regulations, this data also is incomplete.

Occupational Safety and Health Administration (OSHA) Federal agency tasked with the responsibility for the occupational safety of employees.

National Institute for Occupational Safety and Health

Beginning in fiscal year 1998, the **National Institute for Occupational Safety and Health (NIOSH)** began the NIOSH Fire Fighter Fatality Investigation and Prevention Program to conduct independent investigations of firefighter line-of-duty deaths to formulate recommendations for preventing future deaths and injuries. The program does not seek to determine fault or place blame on fire departments or individual firefighters, but it does seek to learn from these tragic events and prevent future similar events. The goals of the program are to better define the magnitude and characteristics of line-of-duty deaths among firefighters, develop recommendations for the prevention of deaths and injuries, and to disseminate prevention strategies to the fire service. Another component of the program is the research database containing information for each injury incident. The database serves as a valuable tool for identifying trends and analyzing risk factors among line-of-duty injury deaths. Used in conjunction with individual incident reports, the database helps provide valuable information for developing broad-based recommendations for firefighter injury prevention programs. As with the investigation reports, personal and fire department identifiers are not included in the database.

One of the important goals for the NIOSH Fire Fighter Fatality Investigation and Prevention Program is disseminating investigative reports and other related publications to fire departments, firefighters, program planners, and researchers who can take action to help prevent firefighter line-of-duty deaths and injuries.

National Institute for Occupational Safety and Health (NIOSH) Federal agency responsible for conducting research and making recommendations for the prevention of work-related injury and illness; NIOSH is part of the Centers for Disease Control and Prevention (CDC).

REVIEW OF NATIONAL INJURY STATISTICS

To provide readers of this text with benchmarks and an insight into the type of information that can be obtained from an annual injury report, we examine the 2010 NFPA Firefighter Injury report. The NFPA report is usually published late in the year following the year studied. For example, the 2010 injury report would be published in late 2011. As described previously, this report is a statistical prediction of the national injury experience. The report examines the injury experience from a number of perspectives shown in the following lists. Each report is useful in analyzing and comparing the local problem with the national problem.

Injuries by Type of Duty

The classification is divided into five subcategories (see **Figure 1-4**):

1. Responding/returning
2. Fireground

FIGURE 1-4
Firefighter injuries by type of duty, 2010. Note: If a firefighter sustained multiple injuries for the same incident, only the nature of the single most serious injury was tabulated.

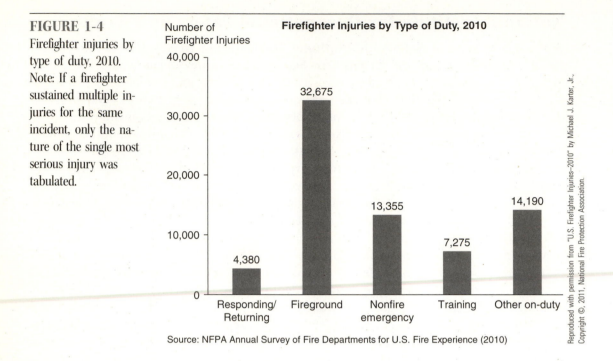

Firefighter Injuries by Type of Duty, 2010

Source: NFPA Annual Survey of Fire Departments for U.S. Fire Experience (2010)

3. Nonfire emergencies
4. Training
5. Other on-duty

In 2010, as in all other years, fireground injuries accounted for the highest number of injuries.

Nature of Injuries

Nature of injury (see **Figure 1-5**) is subdivided into ten classifications, which are reported by type of duty as well.

1. Burns (fire or chemical)
2. Smoke or gas inhalation
3. Other respiratory distress
4. Burn and smoke inhalation
5. Wound, cut, bleeding, bruise
6. Dislocation, fracture
7. Heart attack or stroke
8. Strain, sprain, muscular pain
9. Thermal stress (frostbite, heat exhaustion)
10. Other

Firefighter Injuries by Nature of Injury and Type of Duty, 2010

Nature of Injury	Responding to or Returning from an Incident		Fireground		Nonfire Emergency		Training		Other On-Duty		Total	
	Number	Percent	Number	Percent	Number	Percent	Number	Percent	Number	Percent	Number	Percent
Burns (Fire or Chemical)	50	1.1	1,940	5.9	90	0.7	205	2.8	300	2.1	2,585	3.6
Smoke or Gas Inhalation	20	0.5	1,220	3.7	115	0.9	50	0.7	95	0.7	1,500	2.1
Other Respiratory Distress	60	1.4	440	1.4	110	0.8	130	1.8	200	1.4	940	1.3
Burns and Smoke Inhalation	0	0	555	1.7	10	0.1	35	0.5	35	0.3	635	0.9
Wound, Cut, Bleeding, Bruise	655	15.0	4,650	14.2	1,845	13.8	1,320	18.1	2,640	18.6	11,110	15.5
Dislocation, Fracture	230	5.3	855	2.6	195	1.5	235	3.2	305	2.2	1,820	2.5
Heart Attack or Stroke	70	1.6	175	0.5	100	0.8	135	1.9	330	2.3	810	1.1
Strain, Sprain, Muscular Pain	2,705	61.8	17,250	52.8	8,650	64.8	4,255	58.5	7,525	53.0	40,385	56.2
Thermal Stress (frostbite, heat exhaustion)	205	4.7	2,350	7.2	140	1.1	380	5.2	120	0.9	3,195	4.5
Other	385	8.8	3,240	9.9	2,100	15.7	530	7.3	2,640	18.6	8,895	12.4
	4,380	100.0	32,675	100.0	13,355	100.0	7,275	100.0	14,190	100.0	71,875	100.0

Source: NFPA Survey of Fire Departments for U.S. Fire Experience, 2010
Note: If a firefighter sustained multiple injuries for the percent incident, only the nature of the single most serious injury was tabulated.

FIGURE 1-5
Nature of injuries by type of duty, 2010.

FIGURE 1-6
Fireground injuries by
cause, 2010.

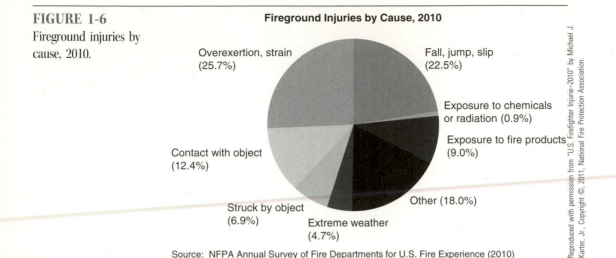

Fireground Injuries by Cause, 2010

Overexertion, strain (25.7%)

Fall, jump, slip (22.5%)

Exposure to chemicals or radiation (0.9%)

Exposure to fire products (9.0%)

Contact with object (12.4%)

Other (18.0%)

Struck by object (6.9%)

Extreme weather (4.7%)

Source: NFPA Annual Survey of Fire Departments for U.S. Fire Experience (2010)

Fireground Injuries by Cause

As fireground injuries historically cause the highest number of injuries per type of duty, fireground injuries by cause are also reported (see **Figure 1-6**). The eight categories for fireground injuries by cause are:

1. Struck by object
2. Stepped on, contact with object
3. Extreme weather
4. Exposure to fire products
5. Exposure to chemicals or radiation
6. Fell, slipped, jumped
7. Overexertion
8. Other

Average Number of Fires and Fireground Injuries per Department by Population

The average number of fires and fireground injuries per department by population for a community protected in 2010 is displayed in **Figure 1-7**. These results show that the number of fires a fire department responds to is directly related to the population protected, and that the number of fireground injuries incurred by a department is directly related to its exposure to fires. The chart also shows the number of injuries per 100 firefighters.

**Average Number of Fires, Fireground
Injuries and Injury Rates
by Population of Community Protected, 2010**

Population of Community Protected	Average Number of Fires	Average Number of Fireground Injuries	Number of Fireground Injuries per 100 Fires	Number of Fireground Injuries per 100 Firefighters
500,000 to 999,999	2,638.7	81.0	3.1	7.2
250,000 to 499,999	1,020.4	28.5	2.8	6.4
100,000 to 249,999	511.7	12.0	2.3	5.3
50,000 to 99,999	218.2	4.4	2.0	4.2
25,000 to 49,999	117.8	2.2	1.9	3.4
10,000 to 24,999	62.5	1.0	1.6	2.4
5,000 to 9,999	36.1	0.6	1.7	1.7
2,500 to 4,999	22.0	0.3	1.4	1.3
Under 2,500	10.6	0.2	1.9	0.9

Source: NFPA Survey of Fire Departments for U.S. Fire Experience, 2010

FIGURE 1-7

Average number of fires, fireground injuries, and injury rates by population of community protected, 2010.

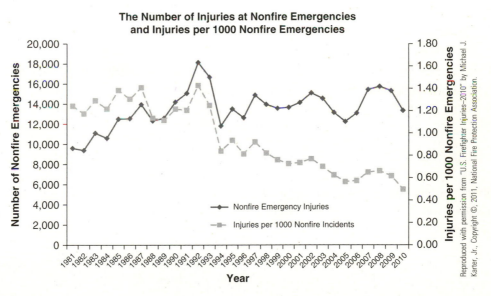

FIGURE 1-8

Number of injuries at nonfire emergencies and injuries per 1000 nonfire emergencies.

The NFPA report further summarizes the data presented in Figures 1-4 through 1-8 in the text of the report. The report also provides an overview and lengthy reports on selected incidents.

NOTE Review of this annual report must be required for safety and health program managers.

WHAT IS BEING DONE

Firefighter Life Safety Initiatives

An unprecedented gathering of the leadership of the American fire service occurred on March 10–11, 2004, when more than 200 individuals assembled in Tampa, Florida to focus on the troubling question of how to prevent line-of-duty deaths. The first-ever National Fire Fighter Life Safety Summit was convened to bring the leadership of the fire service together for two days to focus all of their attention on this one critical concern. Every identifiable segment of the fire service was represented and participated in the process.

The National Fallen Firefighters Foundation (http://www.everyonegoeshome.com) hosted the summit as the first step in a major campaign. In cooperation with the United States Fire Administration, the Foundation has established the objectives of reducing the fatality rate by 25% within 5 years and by 50% within 10 years. The purpose of the summit was to produce an agenda of initiatives that must be addressed to reach those milestones and to gain the commitment of the fire service leadership to support and work toward their accomplishment.

The summit itself marked a significant milestone, because it was the first time that a major gathering had been organized to unite all segments of the fire service behind the common goal of reducing firefighter deaths. It provided an opportunity for all of the participants to focus on the problems, jointly identify the most important issues, agree upon a set of key initiatives, and develop the commitments and coalitions that are essential to moving forward with their implementation. Since this initial meeting, there have been several other meetings that further refined the recommendations from this summit.

The result of this work has been the adoption of the 16 Firefighter Life Safety Initiatives (see **Figure 1-9**). Educational materials and other resources have been developed to support this mission. These are available on the National Fallen Firefighters Foundation website, www.everyonegoeshome.com. The report from this summit can be found in Appendix C.

As described previously, a number of changes have occurred since the publication of NFPA 1500 and the development of the 16 Firefighter Life Safety Initiatives (FLSI). Departments are taking a more proactive approach to safety and health. **Risk management** is

risk management A process of using the available resources of an organization to plan and direct the activities in the organization so that detrimental effects can be minimized.

16 FIREFIGHTER LIFE SAFETY INITIATIVES

1. Define and advocate the need for a cultural change within the fire service relating to safety; incorporating leadership, management, supervision, accountability and personal responsibility.

2. Enhance the personal and organizational accountability for health and safety throughout the fire service.

3. Focus greater attention on the integration of risk management with incident management at all levels, including strategic, tactical, and planning responsibilities.

4. All firefighters must be empowered to stop unsafe practices.

5. Develop and implement national standards for training, qualifications, and certification (including regular recertification) that are equally applicable to all firefighters based on the duties they are expected to perform.

6. Develop and implement national medical and physical fitness standards that are equally applicable to all firefighters, based on the duties they are expected to perform.

7. Create a national research agenda and data collection system that relates to the initiatives.

8. Utilize available technology wherever it can produce higher levels of health and safety.

9. Thoroughly investigate all firefighter fatalities, injuries, and near misses.

10. Grant programs should support the implementation of safe practices and/or mandate safe practices as an eligibility requirement.

11. National standards for emergency response policies and procedures should be developed and championed.

12. National protocols for response to violent incidents should be developed and championed.

13. Firefighters and their families must have access to counseling and psychological support.

14. Public education must receive more resources and be championed as a critical fire and life safety program.

15. Advocacy must be strengthened for the enforcement of codes and the installation of home fire sprinklers.

16. Safety must be a primary consideration in the design of apparatus and equipment.

LEARN MORE: www.EveryoneGoesHome.com

Funding provided by the Department of Homeland Security, Assistance to Firefighters Grant and the generosity of Fireman's Fund Insurance Company.

 Homeland Security

 Fireman's Fund Insurance Company

Courtesy of the National Fallen Firefighters Foundation

FIGURE 1-9
16 Firefighter Life Safety Initiatives.

now a concept as common in the fire station as it is in private industry. What has been and is being done can probably fit nicely into the acronym SAFEOPS, which means:

- *Supervision.* Some departments have hired risk management specialists, assigned safety officers, empowered supervisors to be more safety aware, hired or contracted with fitness and wellness experts, and in many cases contracted with physicians to handle physicals and medical surveillance of department employees. (*FLSI # s 1,2,3,4,5,6,9*)

- *Attitude.* Employees are made aware of the safety and health problems and empowered to be proactive concerning safety. When training begins, there is an emphasis placed on safety; in national journals, there are articles relating to safety and health almost monthly. By virtue of the resources put forth and the supervision focusing on safety and health, employees' attitudes toward safety are slowly improving. Attitude is the key to the success of any safety program—we must constantly seek improvement in attitude. (*FLSI # s 1,3,14*)

- *Fitness/wellness.* Many departments have increased their commitment to fitness and wellness. Fire stations are equipped with workout equipment (see **Figure 1-10**), and in some cases departments are contracting with local gyms. Mental fitness has become a priority, with departments having critical-incident stress debriefing teams and employee assistance programs, including access to these programs for families. Some departments offer classes on weight reduction, smoking cessation, and eating right. Fitness and wellness are critical issues in the overall safety and health program. (*FLSI # s 5,6,13*)

FIGURE 1-10
Fitness equipment should be provided in the station.

© Cengage Learning 2013

- *Education.* The level of safety education, both career and volunteer, for both entry-level and supervisory positions has increased, providing for a broad professional knowledge base that improves the ability to collect, analyze, and present data related to problems. Programs have been developed to better prepare employees to safely perform their jobs, such as how to lift correctly to prevent sprains and strains. (*FLSI # s 1,3,5,7,8,9*)

- *Organizational involvement.* Employee-based safety and health committees, fitness committees, and labor/management relations committees support the safety and health programs by developing new procedures or analyzing new equipment. Employee suggestion programs for safety-related improvements allow all members of an organization to provide input and be stakeholders in making their jobs safer. (*FLSI # s 1,2,5,8,9,11,12,14,15,16*)

- *Procedures.* Operational procedures have been adopted or changed by many departments to reflect the needs of safety and health programs. Included are incident management systems, personnel accountability procedures, use of personal protective equipment, and responding to nonemergency incidents in a nonemergency mode. (*FLSI # s 3,4,11,12*)

- *Standards/regulations.* Standards and regulations have also changed to meet the needs of the occupation. **NFPA 1403,** *Standard on Live Fire Training Evolutions* was written in hopes of preventing future deaths and injuries related to live fire training. Federal agencies also have promulgated regulations aimed at reducing occupational injuries and deaths. The OSHA **bloodborne pathogen** regulation (CFR 1910-1030) requires procedures and precautions to prevent the spread of bloodborne diseases. Many departments may have adopted these procedures even if they had not been forced to comply. (*FLSI # s 5,6,11,12,15,16*)

This is just an overview of initiatives and programs that have been developed to improve the safety and health of emergency responders. Many are not universally accepted or practiced. Hopefully, this text will provide readers with further insight into some of these programs and procedures and motivate them to seek implementation in their own departments.

| **NFPA 1403** | National Fire Protection Association's consensus standard on live fire training evolutions. |
| **bloodborne pathogen** | Disease carried in blood or blood products. |

IS IT WORKING?

Even with the changes and emphasis on emergency service occupational safety and health, it is apparent by the statistics that the rate of illness, injuries, and fatalities has remained relatively constant. There are still a number of firefighter fatalities each year during training exercises. Both the fire service and EMS systems experience responder injuries and deaths during the responding and returning phase of the incident. Often these involve responders in their privately owned vehicles.

Does this mean that the safety and health efforts should be halted? Certainly not; in fact, it points out that the efforts must be continued and expanded. Every agency must make safety and health a top organizational priority. Solid research must be conducted to find out what is working and to develop new strategies to improve the safety and health of responders. This work has already begun.

One example occurred in Phoenix, Arizona, after the fire department experienced a line-of-duty death. While the department had long before developed procedures for the assignment of rapid intervention teams, is was not until after this tragic event that the effectiveness of these teams was evaluated. Phoenix performed 200 rapid intervention drills as part of the recovery process following this fireground death. The results of the study were widely published and in many cases surprised fire service leaders. The results showed that rapid intervention is not rapid; in fact, in the drills, it took rescue crews an average of 2.50 minutes to get prepared. From the firefighter distress signal to the rapid intervention team entry was 3.03 minutes followed by 5.82 minutes for them to make contact with the downed firefighter. The average total time inside the building for each rapid intervention team was 12.33 minutes, with an average total time for rescue of 21 minutes. It was also found that it takes 12 firefighters to rescue 1, that 1 in 5 rapid intervention team members will get into some type of trouble themselves, and that a 3000-psi self-contained breathing apparatus (SCBA) bottle has 18.7 minutes of air (plus or minus 30%).

These kinds of hands-on studies are necessary to answer the question, "Is it working?" or, more importantly, "How well is it working?" New programs are being developed frequently to increase preparedness and to minimize the potential for injury and deaths among responders. Many of the concepts developed are presented in this text.

SUMMARY

Historically, prior to the publication of *America Burning*, little attention was given to overall safety and health in this profession. Occupational safety and health in industry can be traced back hundreds of years. However, the emphasis is relatively new for emergency service providers.

Emergency responders are in a hazardous profession. It is unlikely given the nature of the occupation that all injuries and deaths could be totally eliminated. However, with good risk management the frequency and severity of injuries could be reduced. The first step in designing a safety and health program is to identify and understand the problem. This can be accomplished through analysis of local data and comparisons with the national experience. Several organizations provide national sources for data relating to safety and health programs. The National Fire Protection Association publishes an annual report of firefighter injuries and firefighter deaths, the United States Fire Administration compiles injury and fatality data through the use of the National Fire Incident Reporting System, the International Association of Fire Fighters produces an injury and death report annually, and the Occupational Health and Safety Administration collects and analyzes data relating to occupational injuries. The National Institute for Occupational Safety and Health investigates firefighter line-of-duty deaths. All of these resources should be used to analyze safety and health problems.

Many changes have occurred in the last 20 to 30 years regarding emergency responder safety, health, and wellness. In the last decade, an unprecedented meeting of 200 fire service leaders in Tampa resulting in the development of the 16 Firefighter Life Safety Initiatives.

These changes include increased awareness and placing greater priority on safety issues by front line supervisors, improvement in safety attitudes at all levels of the organization, increased emphasis on fitness and wellness, higher levels of education for entry-level employees and supervisors, involvement of all levels of the organization through committee and suggestion programs, revision and adoption of procedures related to safety operations, and promulgation of standards and regulations designed to improve the safety and health of responders. Even with all of these changes, there is still significant progress that could be made. Through research, the development of new procedures and equipment will help to make the job of emergency service responder safer.

Concluding Thought: Improvements have been made to improve responders' safety and health. However, as represented in this chapter, (1) the number of fires is down, (2) EMS incidents remain on the increase, but (3) the injuries have remained constant. Therefore, we need better programs, research, data, and organizational commitment to improve even more. Each organization can and should benchmark itself against similar organizations to further improve the statistics.

REVIEW QUESTIONS

1. What do the letters of the acronym SAFEOPS stand for?
2. Which of the following organizations uses a sampling survey of every type of fire department, both paid and volunteer, in every state for injury and death statistics?
   a. USFA
   b. IAFF
   c. NFPA
   d. OSHA
   e. None of the above survey every type of department
3. According to the preceding text, occupational safety and health moved ahead rapidly during which of the following decades?
   a. 1950s
   b. 1960s

    c. 1970s

    d. 1980s

4. Which major initiative did the IAFF and the International Association of Fire Chiefs work on together?

    a. NFPA 1500

    b. OSHA 1910.120

    c. The SAFEOPS approach

    d. A fitness/wellness program

Use Figures 1-4 through 1-8 to answer questions 5 through 10.

5. How many injuries per 1000 non-fire emergencies occurred in 2010?

6. What percentage of injuries occurred on the fireground?

7. The most sprains and strains occurred during what type of duty?

8. What population range has the highest rate of injuries?

9. What population range has the highest rate of fires?

10. What caused the greatest percentage of fireground injuries?

ACTIVITIES

1. Obtain a copy of the most recent death or injury report from any of the organizations in this chapter. Review the statistics and compare them to those of your department.

2. Using the SAFEOPS approach, analyze what your department has done since the publication of the 1987 edition of *NFPA 1500* in terms of improving responder safety and health. List these approaches and evaluate their effectiveness.

3. List the 16 Firefighter Life Safety Initiatives. Which are in place in your department?

NOTES

1. Ludwig, Gary. *On Scene*, May 8, 1998. International Association of Fire Chiefs.

2. National Commission on Fire Prevention and Control, 1973. Washington, D.C.

CHAPTER 2

Review of Safety-Related Regulations and Standards

LEARNING OBJECTIVES

Upon completion of this chapter, you should be able to:

- Discuss the difference between regulations and standards.
- Discuss the concept of standard of care.
- List and discuss federal regulations that have an impact on emergency responder safety and health programs.
- Describe the OSHA General Duty Clause.
- List and discuss the major National Fire Protection Association standards that have an impact on emergency responder occupational safety and health programs.
- Discuss the role of related regulations and standards and their safety and health implications.

On February 9, 2007, a probationary fire fighter died while participating in a live-fire training evolution at an acquired structure. The victim's class was conducting a live-fire training drill that is required by the department's training protocol for their *NFPA 1001, Standard for Fire Fighter Professional Qualifications*, Fire Fighter I. The victim was part of a four-person engine company, led by an adjunct instructor, that made the initial attack on a training fire in a vacant, condemned, three-story, end-unit townhouse. The scenario called for the victim's crew to enter the front of the townhouse and proceed to the third floor to find and extinguish any fire on the third floor. They were to bypass any fire on the second floor so that the second due engine could practice suppression on that floor. The victim's crew encountered heavy fire on the second floor and third floor stairwell as they proceeded to the third floor. The victim, operating the nozzle, and the adjunct instructor attempted to fight fire on the third floor, but conditions made it untenable. The adjunct instructor was able to exit through a window located on the third floor landing followed by a fire fighter who was backing up the victim on the hoseline. However, the victim got stuck attempting to exit the window, which was 41 inches above the floor. The victim became unresponsive as the adjunct instructor and other fire fighters attempted to free her from the window. After she had been freed, she was transported to a local trauma center where she was pronounced dead.

Recommendations by the NIOSH investigators included the following:

1. Fire departments should conduct live-fire training exercises in accordance with the most recent edition of *NFPA 1403, Standard on Live-Fire Training Evolutions.*

2. Fire departments should ensure all training and education, including live fire training, is conducted under the direct supervision of a qualified instructor(s) who meets the requirements of *NFPA 1041, Standard for Fire Service Instructor Professional Qualifications.*

3. Fire departments should provide the Training Academy and Safety Division with adequate resources, personnel, and equipment to accomplish their training mission safely.

4. Fire departments should screen recruits to ensure they meet the physical performance requirements as established by the fire department prior to entering a training program to become a fire fighter.

5. Fire departments should develop and maintain a comprehensive respiratory protection program which complies with *NFPA 1404, Standard for Fire Service Respiratory Protection Training.*

6. Fire departments should ensure all recruits meet the requirements of *NFPA 1582, Standard on Comprehensive Occupational Medical Program for Fire Departments* prior to entering the training program.

7. Fire departments should develop inspection criteria to ensure that all protective ensembles meet the requirements of *NFPA 1851, Standard on Selection, Care, and Maintenance of Structural Fire Fighting Protective Ensembles.*

8. Fire departments should ensure coordinated communication between the Instructor-in-charge and the live-fire training participants.

9. Fire departments should utilize the incident command system and a personnel accountability system, for all incidents, including live-fire training exercises, that meets the requirements of *NFPA 1561, Standard on Emergency Services Incident Management System.*

10. Fire departments should create a training atmosphere that is free from intimidation and conducive to learning.

11. States should develop a permitting procedure for live-fire training to be conducted at acquired structures and also ensure that all the requirements of *NFPA 1403* have been met before issuing the permit.

Of the 11 recommendations provided as a result of the investigation and analysis, seven are related to nationally recognized standards. This fact underscores the need for the emergency service Health and Safety Officer to have an in-depth knowledge and understanding of the regulations and standards as well as the application of these to the operations of the organization.

Source: Information for this case study was taken in part from CDC/NIOSH Fire Fighter Fatality Investigation Report F2007-09. http://www.cdc.gov/niosh/fire/reports/face200709.html

---

INTRODUCTION

As emphasized in the case study, the knowledge and understanding of health and safety–related standards and regulations is essential in all emergency service functions. "We have not adopted that standard" or "We are not an OSHA state" are not necessarily valid reasons to disregard the existence of standards and regulations.

In reviewing the 16 Firefighter Life Safety Initiatives, it is clear that six are directly or closely related to regulations and standards, enforcing the notion that regulations and standards are essential components of, and play an important role in, a comprehensive safety and health program. These six Initiatives are:

5. *Develop and implement national standards for training, qualifications, and certification (including regular recertification) that are equally applicable to all firefighters based on the duties they are expected to perform.*

6. *Develop and implement national medical and physical fitness standards that are equally applicable to all firefighters, based on the duties they are expected to perform.*

11. *National standards for emergency response policies and procedures should be developed and championed.*

12. *National protocols for response to violent incidents should be developed and championed.*

15. *Advocacy must be strengthened for the enforcement of codes and the installation of home fire sprinklers.*

16. *Safety must be a primary consideration in the design of apparatus and equipment.*

REGULATIONS VERSUS STANDARDS

There are differences between **regulations** and **standards**. Regulations are promulgated at some level of government by a governmental agency and have the force of law. Standards, sometimes called *consensus standards*, do not have the weight of law unless they are adopted as law by an authority having jurisdiction.

From the safety program management standpoint, an understanding of these applicable regulations and standards is essential and invaluable. These documents can actually provide a road map for an agency's safety and health program. This chapter focuses on the regulations and standards that apply to emergency service safety and health. Emphasis is placed on those that have the greatest effect and impact on the programs, but others are introduced also.

regulations Rules or laws promulgated at the federal, state, or local level with a requirement to comply.

standards Often developed through the consensus process, standards are not mandatory unless adopted by a governmental authority.

Regulations

Regulations carry the weight of law and are mandatory in their requirements based on federal and, in some cases, state and local legislation. The entire collection of federal regulations is contained within the 50 titles of the **Code of Federal Regulations (CFR)**. These mandatory requirements impact on emergency service safety and health programs, primarily those regulations found in Title 29 CFR, which are the OSHA regulations. These specific standards are discussed in further detail later in this chapter in the section Occupational Safety and Health Administration Regulations.

Code of Federal Regulations (CFR) The document that contains all of the federally promulgated regulations for all federal agencies.

NOTE Remember, these regulations have the power of law and are enforced by the federal agency responsible for them.

Some states have adopted federal laws, which then become mandatory state requirements. For example, the State of Florida Division of State Fire Marshal has adopted several OSHA regulations as they apply to firefighters and fire department employers, including Section 1910.120 and portions of Sections 1910.134, 1910.146, and 1910.156 of Part 1910 of the Occupational Safety and Health Standards, 29 Code of Federal Regulations. In cases in which states adopt a regulation, state officials provide the enforcement.

Standards

Other published documents do not mandate compliance. These documents are commonly known as **consensus standards** because a group of professionals with a specific expertise have come together and agreed on how a specific task should be performed. The NFPA is one such standards-making group. The NFPA publishes standards, recommended practices, and codes. NFPA standards are documents in which the text includes the mandatory provisions of the requirements.

NOTE *Recommended practices* are documents that are similar in content to standards and codes, but are non-mandatory in compliance. *Codes* are standards that cover broad subject areas, which can be adopted into law independently of other codes or standards.

The NFPA has no enforcement authority or power; its standards are considered advisory. However, a jurisdiction can adopt an NFPA standard and then the adopting authority would have legal rights to enforce the standard. The most common example of this process is the adoption of *NFPA 101, Life Safety Code*®. Both local and state governments have adopted this standard as part of their fire prevention programs.

NOTE Once a standard is adopted, the adopting governmental agencies have the power for enforcement.

The NFPA has developed a number of standards that impact safety and health programs. The most notable is *NFPA 1500, Standard on Fire Department Occupational Safety and Health Program*. Along with the *NFPA 1500* standard come a number of other standards in the *1500* series, including medical requirements, infection control, and fire department safety officer. Further discussion of the NFPA standards is presented later in this chapter.

consensus standards Standards developed by consensus of industry or subject area experts, which are then published and may or may not be adopted locally. Even if not adopted as law, these standards can often be used as evidence for standard of care.

It is important to remember that these are consensus standards and are not mandatory unless adopted into law by local or state legislation. However, due to the fact that a group of professionals with related interests and expertise have come together and agreed on some minimum level of performance, these standards tend to become a **standard of care** in the particular subject.

standard of care The concept of what a reasonable person with similar training and equipment would do in a similar situation.

STANDARD OF CARE

The concept of standard of care is well known in the emergency medical field. From the first day of training, this concept is introduced to make students realize that to avoid liability they must perform in the same way as another reasonable person with the same training and equipment would perform.

NOTE The concept of standard of care is simple—everyone has certain expectations when it comes to performance.

When a deposit is made to your checking account at the bank, you have a reasonable expectation that the teller will put the money into the right account. If mistakes are made, the bank has to take responsibility for the teller's actions.

This same concept can be applied to safety and health issues and closely relates to the existence of standards. The publication of a safety standard in and of itself has an impact on standard of care. For example, *NFPA 1500* requires that all responders to hazardous situations be equipped with a **personal alert safety system (PASS)**. Should a firefighter get lost in a building fire and die, the Fire Chief or Health and Safety Officer may find themselves in court answering the question of why a PASS device was not provided. But, you say, NFPA is not law and has not been legally adopted in my jurisdiction. This answer will not be a viable defense when the firefighter's family sues for not following reasonable industry standards, because the standard of care has been defined by the NFPA document.

Standard of care is not a static concept but instead is very dynamic. It changes with new technologies and the development of regulations, standards, and procedures. Twenty-five years ago it was an acceptable standard of care to allow firefighters to fight fires without PASS devices. They did not exist and, if they did, no published document required them; therefore, it was an acceptable practice to enter a building without a PASS.

personal alert safety system (PASS) A device that produces a high-pitched audible alarm when the wearer becomes motionless for some period of time; useful to attract rescuers to a downed firefighter.

TABLE 2-1 States and Territories with OSHA Plans

Alaska	Indiana	Minnesota	North Carolina	Utah
Arizona	Iowa	Nevada	Oregon	Vermont
California	Kentucky	New Mexico	Puerto Rico	Virgin Islands
Connecticut	Maryland	New Jersey	South Carolina	Virginia
Hawaii	Michigan	New York	Tennessee	Washington
Illinois				Wyoming

(Please note that the Connecticut, Illinois, New Jersey, New York, and Virgin Islands plans cover public sector employment only.)
Source: http://www.osha.gov.

© Cengage Learning 2013

OCCUPATIONAL SAFETY AND HEALTH ADMINISTRATION REGULATIONS

A number of OSHA regulations affect emergency service safety and health programs. Those most commonly used include confined space operations, respiratory protection, hazardous waste operations, and those regarding bloodborne pathogens. The following paragraphs describe each of these regulations in terms of applicability to safety and health programs, and review the contents of each.

The application of OSHA regulations to public employers is a confusing issue. Under the Occupational Safety and Health Act of 1970, federal OSHA has no direct enforcement authority over state and local governments. However, a state may opt to implement its own enforcement program and may do so as long as federal OSHA approves the state's safety and health plan. A number of requirements, found in Section 18 of the Occupational Safety and Health Act, must be met for a state to have an approved plan. Some of the requirements include: that the state be willing to assume responsibility for development and enforcement of occupational safety and health standards; assurances that the state will provide adequate funds for the administration and enforcement of the standards; that the state will, to the extent permitted by its law, establish and maintain an effective and comprehensive occupational safety and health program applicable to all employees of public agencies of the state and its political subdivisions; and that the state make the same reports to OSHA as it would if the state plan were not in effect.

There are currently twenty-seven states or territories that have state OSHA plans (see **Table 2-1**). In these states, firefighters—including state, county, and municipal—are covered by regulations promulgated by federal OSHA.

The full text of the regulations summarized below can be found at http://www.osha.gov

OSHA 1910.146 (29 CFR 1910.146) Permit-Required Confined Spaces

This regulation contains requirements for practices and procedures to protect employees in general industry from the hazards of entry into permit-required confined spaces. This section does not apply to agriculture, to construction, or to shipyard employment.

FIGURE 2-1
Confined space entry is governed by OSHA regulations.

© Cengage Learning 2013

There are several areas that apply to safety practices and procedures. The confined space regulation defines a confined space as any area that (1) is large enough and so configured that a person can bodily enter and perform assigned work; (2) has limited or restricted means for entry or exit (for example, tanks, vessels, silos, storage bins, hoppers, vaults, and pits are spaces that may have limited means of entry); and (3) is not designed for continuous employee occupancy (see **Figure 2-1**).

Clearly, there are cases when the emergency service employee would be required to enter the environments just defined. When private employers choose to use the local emergency service as their rescue team, 29 CFR 1910.146 (k) requires performance evaluation. Guidance for private employers who wish to use local rescue services can be found in 29 CFR 1910.146 Appendix F. Requirements contained in 1910.146 include those for written plans, atmosphere monitoring, notification, and equipment. Some companies that have permitted confined spaces on site have contracted with the local fire or rescue agency to perform services if needed. Even in the absence of this pre-event agreement, we can expect that when an emergency occurs, the local emergency services will be summoned by a 9-1-1 phone call.

NOTE From a safety program standpoint, these operations should not be performed until written procedures have been developed, the necessary equipment obtained, and the personnel have been properly trained.

OSHA 1910.134 (29 CFR 1910.134) Respiratory Protection

This regulation provides for the control of those occupational diseases caused by breathing air contaminated with harmful dusts, fogs, fumes, mists, gases, smokes, sprays, or vapors; the primary objective is to prevent atmospheric contamination. This shall be accomplished as far as feasible by accepted engineering control measures (for example, enclosure or confinement of the operation, general and local ventilation, and substitution of less toxic materials). When effective engineering controls are not feasible, or while they are being instituted, appropriate respirators shall be used pursuant to this section. A respirator shall be provided to each employee when such equipment is necessary to protect the health of such employee. The employer shall provide respirators that are applicable and suitable for the purpose intended. The employer shall be responsible for the establishment and maintenance of a respiratory protection program, which shall include the requirements in the regulation.

Employers therefore must provide respirators when such equipment is necessary to protect the health of the employee (including **self-contained breathing apparatus (SCBA)** (see **Figure 2-2**). The employer is required to provide respirators that are

self-contained breathing apparatus (SCBA) An atmosphere-supplying respirator for which the breathing air source is designed to be carried by the user.

FIGURE 2-2
SCBA usage and training falls under OSHA's respiratory protection regulation.

applicable and suitable for the purpose intended. The employer is responsible for the establishment and maintenance of a respiratory protection program. The employee shall use the provided respiratory protection in accordance with the instructions and training received. Requirements for a respiratory protection program include the following:

- Procedures for selecting respirators for use in the workplace
- Medical evaluations of employees required to use respirators
- Fit-testing procedures for tight-fitting respirators
- Procedures for proper use of respirators in routine and reasonably foreseeable emergency situations
- Procedures and schedules for cleaning, disinfecting, storing, inspecting, repairing, discarding, and otherwise maintaining respirators
- Procedures to ensure adequate air quality, quantity, and flow of breathing air for atmosphere-supplying respirators
- Training of employees in the respiratory hazards to which they are potentially exposed during routine and emergency situations
- Training of employees in the proper use of respirators, including putting on and removing them, any limitations on their use, and their maintenance
- Procedures for regularly evaluating the effectiveness of the program

Most of these requirements should be incorporated into the organization's **standard operating procedures (SOPs)**, also known as standard operating guidelines (SOGs). Immediately, even from these minimum requirements, one can see the safety program implications. One requirement in the respiratory protection regulation deals with fire scene staffing requirements. The procedures for operations in **immediately dangerous to life and health (IDLH)** environments are found in section 1910.134(g)(3) and further requirements for **interior structural firefighting** are described in 1910.134(g)(4). Summaries of the requirements of these two sections are listed in **Text Box 2-1**.

The requirements of 1910.134(g)(3) and 1910.134(g)(4) have come to be more commonly known as the two-in/two-out rule. Although the regulation does not preclude the firefighters from attempting a rescue in a known life-and-death situation prior to the

standard operating procedures (SOPs) Sometimes called standard operating guidelines, these are department-specific operational procedures, policies, and rules made to assist with standardized actions at various situations.

immediately dangerous to life and health (IDLH) Used by several OSHA regulations to describe a process or an event that could produce loss of life or serious injury if a responder is exposed to or operates in the environment.

interior structural firefighting The physical activity of fire suppression, rescue, or both, inside buildings or enclosed structures that are involved in a fire beyond the incipient stage.

TEXT BOX 2-1

REQUIREMENTS OF OSHA 1910.134(g)(3) AND 1910.134(g)(4)

For all IDLH atmospheres, 1910.134(g)(3) requires that the employer shall ensure the following:

- One employee or, when needed, more than one employee is located outside the IDLH atmosphere.

- Visual, voice, or signal line communication is maintained between the employee(s) in the IDLH atmosphere and the employee(s) located outside the IDLH atmosphere.

- The employee(s) located outside the IDLH atmosphere are trained and equipped to provide effective emergency rescue.

- The employer or designee is notified before the employee(s) located outside the IDLH atmosphere enter the IDLH atmosphere to provide emergency rescue.

- The employer or designee authorized to do so by the employer, once notified, provides necessary assistance appropriate to the situation.

- Employee(s) located outside the IDLH atmospheres are equipped with:

  - Pressure demand or other positive-pressure SCBAs, or a pressure demand or other positive-pressure supplied-air respirator with auxiliary SCBA; and either

  - Appropriate retrieval equipment for removing the employee(s) who enter(s) these hazardous atmospheres where retrieval equipment would contribute to the rescue of the employee(s) and would not increase the overall risk resulting from entry; or

  - Equivalent means for rescue where retrieval equipment is not required under the previous subitem.

In addition to the requirements set forth under paragraph (g)(3), 1910.134(g)(4) requires that for interior structural fires, the employer shall ensure that:

- At least two employees enter the IDLH atmosphere and remain in visual or voice contact with one another at all times.

- At least two employees are located outside the IDLH atmosphere.

- All employees engaged in interior structural firefighting use SCBAs.

Note 1 to paragraph (g): One of the two individuals located outside the IDLH atmosphere may be assigned to an additional role, such as incident commander in charge of the emergency or safety officer, so long as this individual is able to perform assistance or rescue activities without jeopardizing the safety or health of any firefighter working at the incident.

Note 2 to paragraph (g): Nothing in this section is meant to preclude firefighters from performing emergency rescue activities before an entire team has assembled.

Source: http://www.osha.gov

rescue team being in place (although when such a situation occurs, a written report should be completed and forwarded to the fire chief), it does require that before fire-fighters can operate in situations in which SCBA is worn, four firefighters would be required. This ruling has implications to the operations of both career and volunteer fire service agencies because it applies to both. Appendix B contains an excellent and concise document issued by the IAFC and the IAFF that deals with this ruling.

The requirements of 1910.134 should be carefully evaluated by the safety and health officer prior to safety program development. Procedures that address operations when respirators are worn, staffing, the maintenance and repair of respirators, and the training that employees receive must be designed to meet this regulation.

NOTE The OSHA ruling states that when workers are using SCBA in immediately dangerous to life and health (IDLH) environments or unknown environments, the workers shall operate in a buddy system and there shall be at least two persons outside the danger area ready to effect a rescue.

OSHA 1910.120 (29 CFR 1910.120) Hazardous Waste Operations and Emergency Response (HAZWOPER)

The emergency service application of this regulation is clearly described as emergency response operations for releases of, or substantial threats of releases of, hazardous substances without regard to the location of the hazard. This OSHA regulation has a number of health and safety–related requirements, including a written safety and health program for employees involved in hazardous waste operations. This written plan must be made available to all employees and is required to contain, at a minimum, the following sections: (1) a means to identify, evaluate, and control safety and health hazards and provide for emergency response for hazardous waste operations; (2) an organizational structure; (3) a comprehensive work plan; (4) a site-specific safety and health plan, which need not repeat the employer's SOPs; (5) a safety and health training program; (6) a medical surveillance program; (7) the employer's SOPs for safety and health; and (8) any necessary interface between general program and site-specific activities. The regulation also has specific requirements for personal protective equipment, training levels, and the necessary personnel to operate at an applicable incident. As is obvious, the implication of this regulation to emergency services in general is powerful and multifaceted. It is of particular importance to those organizations that provide hazardous materials response. In the case of the training requirements, the employer must certify that the competencies have been met.

The 1910.120 regulation was referenced in the January 1998 OSHA interpretation on staffing, specifically the requirement of personnel to utilize a buddy system (see **Figure 2-3**).

FIGURE 2-3
The HAZWOPER regulations identify minimum requirements for hazardous material operations.

© Cengage Learning 2013

OSHA 1910.156 (29 CFR 1910.156) Fire Brigades

The OSHA fire brigade regulation applies to **fire brigades**, industrial fire departments, private or contractual-type fire departments, and the public fire departments in states that have adopted it, for example, the state of New York. The personal protective clothing requirements of this regulation apply only to those fire brigades that perform interior structural firefighting. The regulation specifically excludes airport crash fire rescue and forest firefighting operations. This regulation contains requirements for the organization, training, and personal protective equipment for fire brigades whenever they are established by employers.

The fire brigade regulation contains sections on (1) the organizational statement (which establishes the existence of the brigade and personnel requirement), (2) training and education, (3) firefighting equipment, (4) protective clothing, and (5) respiratory protection. The requirements for protective clothing reference several applicable National Fire Protection Association standards that will be described later in this chapter.

As with the 1910.134 and 1910.120 regulations, this regulation was also referenced in the January 1998 interpretation on staffing. Specifically, section (f)(1)(i) was cited as it requires that SCBA be provided for all firefighters engaged in interior structural firefighting.

fire brigades The use of trained personnel within a business or at an industry site for firefighting and emergency response.

OSHA 1910.1030 (29 CFR 1910.1030) Occupational Exposure to Bloodborne Pathogens

The intent of this regulation was to eliminate or minimize occupational exposures to hepatitis B virus (HBV), human immunodeficiency virus (HIV), and other bloodborne pathogens. The regulation is based on the premise that these exposures can be minimized or eliminated through a combination of engineering and work practice controls, personal protective clothing and equipment, training, medical surveillance, hepatitis B vaccination, signs and labels, and other provisions. The regulation applies to all employees who may have occupational exposure to blood or other potentially infectious material as defined in the regulation (see **Figure 2-4**).

NOTE This regulation requires employers to have an exposure control plan that is reviewed and updated annually or sooner, as needed, to reflect new procedures or new equipment.

Other requirements of the regulation include exposure determination, engineering and work practice controls, labeling of waste, housekeeping, personal protective

---

FIGURE 2-4
Personal protective equipment is required to limit exposure to bloodborne pathogens.

© Cengage Learning 2013

clothing and equipment, record keeping, and decontamination procedures. The 1910.1030 regulation requires employers to provide hepatitis vaccinations to all employees with occupational exposure risk.

This regulation must be consulted when writing medical/rescue response procedures for the department's safety program, because almost all emergency service employees fall within the scope of this document in terms of occupational exposure.

OSHA General Duty Clause

When the Occupational Safety and Health Act was enacted in 1970, it was clear to Congress that not all safety and health situations could be covered by the regulations. Knowing this, Congress included in the act the General Duty Clause, Section 5(a)(1).

The general duty clause states that each employer shall furnish to each of his employees employment and a place of employment that is free of recognized hazards that are causing or are likely to cause death or serious harm to his employees. Generally the OSHA Review Commission and court precedent have shown that the following elements would be necessary to prove a violation of the general duty clause:

- The employer failed to keep the workplace free of a hazard to which employees of the employer were exposed.
- The hazard was recognized.
- The hazard was causing or was likely to cause death or serious injury.
- There was a feasible and useful method to correct the hazard.

The implication of this clause to the safety and health program comes from a stated relationship with national consensus standards such as the NFPA. A consensus standard can be used to show industry recognition of a hazard. However, the hazard must be recognized in the employers' industry, not an industry other than the employers' industry. The reference of national consensus standards for guidance in enforcement under the General Duty Clause 5(a)(1) can only be used when there is absence of specific OSHA regulation.

NOTE Therefore, for example, in the absence of a specific OSHA regulation, a national consensus standard (such as *NFPA 1500*) can be used as reference for guidance in enforcement of the general duty clause.

This is an important consideration for the safety and health program manager when assessing the need to comply with consensus standards.

NFPA STANDARDS

With the first publication of *NFPA 1500* in 1987, and the movement toward a greater concern for firefighter safety, came the need to establish minimum performance criteria. To fulfill this need, the NFPA continued to develop other safety- and health-related standards in the *1500* series. These standards focused on infection control, fire department safety officers, medical requirements, and incident management systems. Because each of these has an impact on safety program management, each is described in this section.

NFPA 1500, Standard on Fire Department Occupational Safety and Health Program

Clearly the most significant document in the development of the firefighter safety movement was the 1987 adoption of *NFPA 1500*. This standard was adopted after having, at the time, the most comments ever in the history of the NFPA's standards-making process, and it was the first consensus standard for fire service occupational safety and health programs. It was thought that this standard would financially break fire departments and possibly put some out of business completely. The *NFPA 1500* standard is on a five-year revision cycle. As revisions occur, data is analyzed so the standard not only addresses current issues, but addresses them in a manner that takes advantage of current technologies and procedures,

The *NFPA 1500* standard covers the essential elements of a comprehensive safety and health program and incorporates other standards into it as requirements. *NFPA 1500* lists fifty-five other documents that are incorporated by reference, forty-seven of which are other NFPA standards. These referenced documents or portions of them are considered part of the *NFPA 1500* standard. See **Figure 2-5**.

NFPA 1500, as described in the standard's introduction, takes a holistic approach to fire service health and safety and provides the framework for any fire service–related safety and health program, including considerations for EMS response. The requirements outlined in *NFPA 1500* are considered the minimum for an organization providing rescue, fire suppression, emergency medical services, hazardous materials operations, special operations, and related emergency service activities, whether full or part time, as public, governmental, military, private, and industrial fire departments. The standard is not intended to apply to industrial fire brigades, which are covered in *NFPA 600, Standard on Industrial Fire Brigades*. Industrial fire brigades differ from industrial fire departments, in that members of industrial fire departments have the primary responsibility for fire protection, whereas industrial fire brigade members' primary responsibility is elsewhere in the facility, and they only handle fires as they first occur or are in the incipient stage. Again keep in mind that the standard is a minimum standard; nothing is intended to restrict the program manager to go beyond these requirements. The standard requires that adoption of a risk management plan that includes a plan for compliance with *NFPA 1500*.

The many facets of a safety program can be found in the 12 chapters of *NFPA 1500*. The appendices provide valuable guidance to the safety and health program manager. **Text Box 2-2** lists the subject areas found in *NFPA 1500*.

NFPA 1500 provides the user with comprehensive direction for developing a safety and health program. Even if not adopted, this standard is known throughout the

FIGURE 2-5

The *NFPA 1500* standard directly references 45 other NFPA standards.

© Cengage Learning 2013

TEXT BOX 2-2

SAFETY AND HEALTH SUBJECTS IN NFPA 1500

Fire Department Administration

Training, Education, and Professional Development

Fire Apparatus, Equipment, and Driver/Operators

Protective Clothing and Protective Equipment

Emergency Operations

Facility Safety

Medical and Physical Requirements

Member Assistance and Wellness Programs

Critical Incident Stress Programs

emergency field and is accepted as a standard of care. No safety program manager should be without a copy of *NFPA 1500* and its referenced documents.

As with all NFPA standards, *NFPA 1500* is revised on a cycle to ensure that the standard meets current procedures, technologies, and best practices.

NFPA 1521, Standard for Fire Department Safety Officer

With the 1992 edition of the *Standard for Fire Department Safety Officer*, the NFPA changed the standard number from *1501* to *1521* to better fit the numbering plan for fire service occupational safety and health documents. Revisions to *NFPA 1521* continue to make the standard conform to current practices and requirements. This standard contains minimum requirements for the assignment, duties, and responsibilities of both the health and safety officer and the incident safety officer for a fire department or other fire service organization (**Figure 2-6**).

The functions of a fire department's health and safety officer (HSO) are defined in chapter 5 of the standard, and those of the incident safety officer (ISO) in chapter 6. The fire chief appoints the HSO, while the ISO is a qualified member who works directly for the Incident Commander (IC) as part of the command staff. **Table 2-2** compares the various responsibilities and functional areas as required by *NFPA 1521*. The organization's health and safety program manager should meet the requirements of this standard, but need not be a fire department officer, and should ensure that assigned incident safety officers also meet the standard. The standard provides for assistants to these covered positions when warranted by the complexity of the organization or incident.

FIGURE 2-6
A safety officer should be in place throughout the incident.

© Cengage Learning 2013

TABLE 2-2 Various Responsibilities and Functional Areas as Required by *NFPA 1521*

HEALTH AND SAFETY OFFICER	INCIDENT SAFETY OFFICER
Risk management	Incident management system
Laws, codes, and standards	Incident scene safety
Training and education	ISO for fire suppression incidents
Accident prevention	ISO for emergency medical service operations
Accident investigation—procedures and review	ISO for hazardous materials operations
Records management and data analysis	ISO for special operations
Apparatus and equipment for compliance with standards	Accident investigation and review
Facility inspection	Postincident analysis
Health maintenance	
Liaison for safety and health committee	
Occupational safety and health committee	
Infection control	
Critical incident stress management	
Postincident analysis	

© Cengage Learning 2013

NFPA 1561, Standard on Emergency Services Incident Management System

NFPA 1500 requires that fire scene operations be conducted using an Incident Management System (IMS). *NFPA 1561* contains the minimum requirements for an IMS that can be used for all emergency incidents with the purpose of defining and describing the essential elements. To this end, chapter 4 of the standard describes the system implementation, including risk management, system flexibility, resource accountability, incident scene rehabilitation, system qualifications, and training and qualification. The standard is updated to reflect compliance with the National Incident Management System.

Chapter 5 of the standard describes the system components, the command structure, coordination, the incident commander, command post, incident action plan, intelligence, unified command, area command, multi-agency coordination system, supervisory personnel, objectives, conflicting orders, and supervisory awareness. Chapter 5 also outlines the duties and responsibilities of the command staff; the public information officer, liaison officer, safety officer, and the general staff; and the operations section, planning section, logistics section, and finance/administration section.

The safety program manager will find this document very helpful when developing procedures for and training on incident management systems.

NFPA 1581, Standard on Fire Department Infection Control Program

NFPA 1581 was developed to reduce the exposures to infectious diseases by responders in both emergency and nonemergency situations, and is compatible with the guidelines and regulations developed by the Centers for Disease Control and Prevention and the U.S. Department of Health and Human Services that apply to emergency service personnel. The standard defines the minimum requirements for fire department infection control programs. It begins with a description and requirements for the program components including a risk management policy, risk management, training and education, infection control officer, health maintenance, immunization and infectious disease screening, and infectious exposures. Chapter 5 sets forth the requirements for the department's facilities, and chapter 6 describes requirements related to fire department apparatus used to transport patients. Chapter 7 covers emergency medical operations including personnel issues, personal protective equipment, resuscitation equipment, and the handling of sharp objects. Cleaning, disinfecting, and disposal issues are addressed in chapter 8; these include skin washing, disinfectants, emergency medical equipment, clothing, and disposal of material. This standard is of value as a resource to assist in the development of a comprehensive infection control program that is in compliance with the OSHA 1910.1030 regulation.

NFPA 1582, Standard on Comprehensive Occupational Medical Program for Fire Departments

NFPA 1582 was developed to cover the medical requirements necessary for persons engaged in rescue, fire suppression, emergency medical services, hazardous materials mitigation, special operations, and other emergency service functions. The requirements in this standard are intended to apply to both existing employees and to candidates. There are two categories of medical conditions in the standard, A and B. Persons with Category A medical conditions are not allowed to perform emergency operations or training functions. Persons with Category B conditions are to be evaluated on a case-by-case basis with the fire department physician. This standard contains the medical requirements for firefighters whether full or part time, paid or volunteer, and specifies the minimum medical requirements.

The subjects covered in *NFPA 1582* include roles and responsibilities, essential job tasks, medical evaluations of candidates, occupational medical evaluation of current members, annual occupational fitness evaluation of current members, and essential job tasks in specific evaluations of medical conditions. This standard is helpful in determining the frequency of physical exams, in developing criteria for the physical exam, and in providing direction to a department physician as to what conditions may disqualify a person from performing emergency service duties.

NFPA 1583, Standard on Health-Related Fitness Programs for Fire Fighters

The focus during the development of this standard was the intent that a firefighter or other emergency service responder should have a comprehensive health program that includes a fitness component. The standard can be used as a companion document to *NFPA 1582*. This standard can also be used in conjunction with the Joint Labor-Management Wellness Initiative developed by the IAFF and the IAFC, which is discussed later in this text.

The requirements in this standard are the minimum requirements for the development, implementation, and management of a health-related fitness program. Subjects include the duties of the health and fitness coordinator and peer fitness trainers, fitness assessments, exercise and fitness training programs, health promotion education, and data collection. The organization's safety and health officer will find this information useful in adding the fitness and wellness component to the complete safety and health program.

NFPA 1584, Recommended Practice on the Rehabilitation Process for Members during Emergency Operations and Training Exercises

This standard describes preparedness, rehabilitation area characteristics, incident scene and training rehabilitation, and post-incident rehabilitation. *NFPA 1584* covers preincident considerations such as having SOPs in place for rehabilitation and training personnel in heat and cold stress. This standard is helpful in developing rehabilitation procedures. The need for emergency incident rehabilitation is described further in Chapter 5 of this text.

NFPA 1710, Standard for the Organization and Deployment of Fire Suppression Operations, Emergency Medical Operations, and Special Operations to the Public by Career Fire Departments

NFPA 1710 was first published in 2001 after a great deal of controversy and public comment. This standard contains requirements relating to the organization and deployment of fire suppression operations, emergency medical operations, and special operations to the public by career fire departments. One of the purposes stated in this standard is to specify the minimum criteria addressing career public fire suppression operations, emergency medical services, and special operations with regard to the occupational safety and health of fire department members.

Chapter 4 of the standard contains the requirements for the fire department organization. The extensive chapter 5 covers the requirements for fire department services, including fire suppression, emergency medical operations, special operations, airport rescue and firefighting, marine rescue and firefighting, and wildfire incidents. Each of these sections has detailed minimum requirements for staffing, deployment, and operations. Chapter 6 describes the requirements for fire department systems including safety and health, communications, training, preincident planning, and incident management.

This standard not only impacts fire department organization and deployment but also has a number of safety and health program implications.

NFPA 1720, Standard for the Organization and Deployment of Fire Suppression Operations, Emergency Medical Operations, and Special Operations to the Public by Volunteer Fire Departments

Similar in content and structure to *NFPA 1710*, this standard contains minimum requirements relating to the organization and deployment of fire suppression operations, emergency medical operations, and special operations to the public by volunteer fire departments. In contrast to specific requirements in *1710*, *NFPA 1720* considers the uniqueness of the volunteer fire service, the different services they provide, and how they deploy and respond. For this reason, many of the specific requirements are left up to the local authority having jurisdiction to determine.

Chapter 4 of the standard is fire department organization, operations, and deployment, including sections about fire suppression organization and operations, intercommunity organization, emergency medical services, and special operations. Chapter 5, as in *NFPA 1710*, describes the requirements for fire department systems.

Safety and health officers in volunteer organizations should be knowledgeable in this standard and work with administrators to set the minimum standards for deployment and operations as they relate to the safety and health program.

Other NFPA Standards

There are forty-seven other NFPA standards incorporated by reference into the 2007 edition of *NFPA 1500*. Five of these standards—*NFPA 1521, NFPA 1561, NFPA 1581, NFPA 1582*, and *NFPA 1583*—are from the *1500* series and were described in the preceding section. The other 42 are described in this section. They are standards that affect safety and health from other standpoints such as, for example, apparatus specifications, live fire training, and performance requirements for protective clothing. It is important to recognize the existence and inclusion of these standards into *NFPA 1500*.

NFPA 10, Standard for Portable Fire Extinguishers applies to the selection, installation, inspection, maintenance, and testing of portable fire extinguishing equipment.

NFPA 101, *Life Safety Code* identifies the minimum life safety requirements for various occupancies.

NFPA 472, Standard for Professional Competence of Responders to Hazardous Materials Incidents identifies the competencies for first responders at the awareness level, first responders at the operational level, hazardous materials technicians, incident commanders, and off-site specialist employees.

NFPA 473, Standard for Competencies for EMS Personnel Responding to Hazardous Materials Incidents identifies the requirements for basic and advanced life support personnel in the prehospital setting.

NFPA 1001, Standard for Fire Fighter Professional Qualifications identifies performance requirements, specifically the minimum requirements for firefighters.

NFPA 1002, Standard for Fire Apparatus Driver/Operator Professional Qualifications identifies minimum job performance for the firefighter driver/operator of fire department vehicles.

NFPA 1003, Standard for Airport Fire Fighter Professional Qualifications identifies the minimum job performance for the airport firefighter.

NFPA 1006, Standard for Rescue Technician Professional Qualifications establishes the minimum job performance requirements necessary for fire service and other emergency response personnel who perform technical rescue operations.

NFPA 1021, Standard for Fire Officer Professional Qualifications identifies the performance requirements necessary to perform the duties of a fire officer and establishes four levels of progression.

NFPA 1051, Standard for Wildland Fire Fighter Professional Qualifications identifies the minimum job performance requirements for wildland fire duties and responsibilities, including Wildland Firefighter I and II as well as Wildland Fire Officer I and II.

NFPA 1071, Standard for Emergency Vehicle Technician Professional Qualifications identifies the minimum job performance requirements for a person to be considered qualified as an emergency vehicle technician. The standard applies to personnel who are engaged in the inspection, diagnosis, maintenance, repair, and testing of emergency response vehicles.

NFPA 1221, Standard for the Installation, Maintenance, and Use of Emergency Services Communications Systems covers the installation, performance, operation, and maintenance of public emergency service communication systems and facilities.

NFPA 1403, Standard on Live Fire Training Evolutions requires the establishment of procedures for training of fire suppression personnel engaged in structural firefighting under live fire conditions (see **Figure 2-7**).

FIGURE 2-7
NFPA 1403 defines
minimum requirements
for live fire training
exercises.

© Cengage Learning 2013

NFPA 1404, Standard for Fire Service Respiratory Protection Training contains the minimum requirements for the training component of the Respiratory Protection Program as required by *NFPA 1500*.

NFPA 1670, Standard on Operations and Training for Technical Search and Rescue Incidents identifies and establishes levels of functional capability for safely and effectively conducting operations at technical rescue incidents.

NFPA 1851, Standard on Selection, Care, and Maintenance of Structural Fire Fighting Protective Ensembles specifies the minimum selection, care, and maintenance requirements for structural firefighting protective ensembles and the individual ensemble elements, which include coats, trousers, coveralls, helmets, gloves, footwear, and interface components, to be compliant with *NFPA 1971*.

NFPA 1852, Standard on Selection, Care, and Maintenance of Open-Circuit Self-Contained Breathing Apparatus (SCBA) specifies the minimum requirements for the selection, care, and maintenance of open-circuit self-contained breathing apparatus (SCBA) and combination SCBA/supplied air respirators (SAR) that are used for respiratory protection during emergency operations in environments where the atmosphere either is Immediately Dangerous to Life and Health (IDLH), or could become oxygen deficient or IDLH.

NFPA 1901, Standard for Automotive Fire Apparatus applies to all new automotive fire apparatus. The 1996 edition combined *NFPA 1902, 1903*, and *1904* into one standard to cover all types of fire department apparatus as opposed to having a different standard for each.

NFPA 1906, Standard for Wildland Fire Apparatus applies to new automotive fire apparatus designed primarily to support wildland fire suppression operations.

NFPA 1911, Standard for the Inspection, Maintenance, Testing, and Retirement of In-Service Automotive Fire Apparatus The standard covers site, environmental, and equipment rules for pump system performance testing, as well as testing frequency and procedures. Requirements and procedures for testing the aerial device on an apparatus, including aerial ladders, elevating platforms, and water towers. Provisions for testing, servicing, and maintenance of fire apparatus systems and items. *NFPA 1912, Standard for Fire Apparatus Refurbishing* specifies the minimum requirements for the refurbishing of automotive fire apparatus utilized for fire fighting and rescue operations.

NFPA 1925, Standard on Marine Fire-Fighting Vessels provides the minimum requirements for marine firefighting vessels and also provides minimum maintenance and testing requirements.

NFPA 1931, Standard on Design of and Design Verification Tests for Fire Department Ground Ladders applies to the specific requirements for the design of and the design verification test for fire department ground ladders. This standard applies to all new ground ladders intended for use by fire department personnel for firefighting operations and training.

NFPA 1932, Standard on Use, Maintenance, and Service Testing of Fire Department Ground Ladders applies to the specific requirements for the use, maintenance, inspection, and service testing of fire department ground ladders.

NFPA 1936, Standard on Powered Rescue Tool Systems specifies the minimum requirements for the design, performance, testing, and certification of powered rescue tool systems including individual components of spreaders, rams, cutters, combination tools, power units, and power transmission cables, conduit, or hose.

NFPA 1951, Standard on Protective Ensembles for Technical Rescue Incidents specifies the minimum design, performance, testing, and certification requirements for utility technical rescue; rescue and recovery technical rescue; and chemicals, biological agents, and radiological particulate (also known as chemical, biological, radiological, and nuclear [CBRN]) technical rescue protective ensembles for use by emergency services personnel during technical rescue incidents.

NFPA 1961, Standard on Fire Hose applies to the testing of new fire hose, specified as attack hose, occupant use hose, forestry hose, and supply hose.

NFPA 1962, Standard for the Inspection, Care, and Use of Fire Hose, Couplings, and Nozzles and the Service Testing of Fire Hose applies to the care of all types of

FIGURE 2-8

Two types of *NFPA 1971-* compliant foot protection.

fire hose, coupling assemblies, and nozzles while in service, both in use and after use.

NFPA 1964, Standard for Spray Nozzles applies to portable adjustable pattern nozzles intended for general fire department use.

NFPA 1971, Standard on Protective Ensemble for Structural Fire Fighting specifies criteria and test methods for protective clothing designed to protect firefighters against environmental effects during structural firefighting (see **Figure 2-8**).

NFPA 1975, Standard on Station/Work Uniforms for Fire and Emergency Service specifies general requirements, performance requirements, and test methods for materials used in the construction of station/work uniforms to be worn by members of the fire service.

NFPA 1977, Standard on Protective Clothing and Equipment for Wildland Fire Fighting specifies the minimum design, performance, testing, and certification requirements for protective clothing, helmets, gloves, and footwear that are designed to protect firefighters during wildland firefighting operations.

NFPA 1981, Standard on Open-Circuit Self-Contained Breathing Apparatus for Fire and Emergency Services specifies requirements for design, performance, testing, and certification of SCBA used in firefighting, rescue, and other hazardous situations.

NFPA 1982, Standard on Personal Alert Safety Systems (PASS) specifies design, performance, and test methods for personal alert safety systems to be used by firefighters engaged in firefighting, rescue, and other hazardous duties.

NFPA 1983, Standard on Fire Service Life Safety Rope and System Components specifies performance, design, and test methods for life safety rope, harness, and hardware used by the fire service.

NFPA 1989, Standard on Breathing Air Quality for Emergency Services Respiratory Protection specifies the minimum requirements for breathing air quality for fire and emergency services organizations that use atmosphere-supplying respirators.

NFPA 1991, Standard on Vapor-Protective Ensembles for Hazardous Materials Emergencies applies to the requirements for vapor-protective suits and replacement components.

NFPA 1992, Standard on Liquid Splash-Protective Ensembles and Clothing for Hazardous Materials Emergencies applies to the requirements for liquid splash–protective suits and replacement components.

NFPA 1994, Standard on Protective Ensembles for Chemical/Biological Terrorism Incidents specifies the minimum requirements for the design, performance, testing, documentation, and certification of protective ensembles designed to protect fire and emergency services personnel from chemical/biological terrorism agents.

NFPA 1999, Standard on Protective Clothing for Emergency Medical Operations specifies minimum emergency medical clothing designed to protect emergency personnel or patients from exposure to liquidborne pathogens during emergency medical operations.

The following five standards are not specifically referenced by *NFPA 1500* but also impact an organization's safety and health program and therefore deserve review.

NFPA 1041, Standard for Fire Service Instructor Professional Qualifications

NFPA 1402, Guide to Building Fire Service Training Centers

NFPA 1410, Standard on Training for Initial Emergency Scene Operations

NFPA 1451, Standard for a Fire Service Vehicle Operations Training Program

NFPA 1620, Standard for Pre-Incident Planning

OTHER RELATED STANDARDS AND REGULATIONS
American National Standards Institute

The American National Standards Institute (ANSI) is a national voluntary standard-making organization. Because ANSI uses the consensus process, its membership is comprised of representatives from many disciplines with expertise in the area in which the standard is being developed. These representatives may include professional, trade, technical, or consumer organizations; state and federal agencies; and individual companies.

Many of the standards developed by ANSI are in response to requirements in other regulations and standards. There are a number of ANSI standards relating to the performance of firefighter and emergency medical service responder personal protective equipment such as helmets and goggles. For example, ANSI/ISEA 207-2006, High Visibility Public Safety Vests, is specifically designed to increase the safety and visibility of these emergency responders. Other approved consensus standards may become ANSI standards; for example, the ANSI Safety and Health Standards Board has also approved some standards developed by the NFPA. In February 2003, ANSI established the Homeland Security Standards Panel (HSSP). This panel is to catalog, promote, accelerate, and coordinate the timely development of consensus standards within the national and international voluntary standards systems intended to meet identified homeland security needs, and to communicate the existence of such standards appropriately to governmental units and the private sector.

The safety and health officer must research applicable ANSI standards during program development.

ASTM International, formerly known as the American Society for Testing and Materials (ASTM), is a not-for-profit organization that develops standards for materials, products, and services. Using individuals and organizations from various disciplines, the ASTM also is a consensus organization. While much of the ASTM's purpose is directed at consumer protection issues, some impact emergency service safety. The ASTM has generated several standards directed toward emergency medical system operations, including training.

Environmental Protection Agency

The Environmental Protection Agency (EPA) has promulgated standard 40 CFR 311, which mirrors the OSHA Hazardous Waste Operations and Emergency Response document. The implication of this action is most important for organizations that do not have state OSHA plans. In these states, the requirements of OSHA 1910.120 apply to public employers because of this EPA regulation. Therefore a public emergency service organization in a non-OSHA state that responds to, or has the potential to respond to, hazardous substance releases must comply with the OSHA requirements.

Centers for Disease Control and Prevention

Guidelines for the Prevention of Transmission of Human Immunodeficiency Virus and Hepatitis B Virus to Health-Care and Public-Safety Workers was released by the Centers for Disease Control and Prevention (CDC) in 1989. This document provides an overview of the modes of transmission of HIV in the workplace, an assessment of the risk of transmission under various assumptions, principles underlying the control of risk, and specific risk-control recommendations for employers and workers. The document also includes information on medical management of persons who have sustained an exposure in the workplace to these viruses. This document is very useful for a department developing the exposure control plan that is required by OSHA 1910.1030.

More recently the CDC has published information on anthrax and other biochemical agents as well as information for first responders when dealing with radiation emergencies. This information is essential as emergency services organizations become better prepared for terror attacks.

These are just two examples of the information available from the CDC. It has many documents available that deal with responder safety, particularly in the area of emergency medical operations.

National Institute for Occupational Safety and Health

The National Institute for Occupational Safety and Health (NIOSH) is part of the CDC and was created as part of the act that created OSHA. The primary mission of NIOSH is to reduce work-related injuries and illnesses. One function of NIOSH of interest to emergency services is the research conducted through the Firefighter Fatality Investigation and Prevention program. This program is described in detail in Chapter 12.

From a regulation and standards standpoint, NIOSH publishes documents applicable to emergency service responders. Information is available on bloodborne infectious disease, personal protective equipment, guidelines for Fire Service Operations in a Hurricane Response, and traumatic incident stress; and numerous recommended practices are available as part of the Firefighter Fatality Investigation and Prevention Program.

Ryan White Comprehensive AIDS Resources Emergency Act of 1990, Subtitle B

Another document that must be reviewed when managing infection control is the Ryan White Act. This act was enacted in 1990 and has been reauthorized most recently in 2009 for four additional years. This act sets forth the requirements for employers to designate an infection control officer. Further, the act provides for the notification of employees by medical facilities after an exposure takes place. The safety program manager must be aware of the act because the notification procedures, the designated officer, and the follow-up procedures should become part of the department's program.

SUMMARY

Many regulations and standards from a number of sources relate to emergency service occupational safety and health. Regulations are laws and can be enforced by the agency that promulgated them. Standards, however, are developed by the consensus of experts in the subject area and only become enforceable if adopted by a state or local government. Although standards may not be officially adopted, they may be used to establish a standard of care or may be referenced by a regulation or used during litigation proceedings.

Many applicable safety and health regulations come from OSHA. These regulations cover public employees in the 27 states and territories that have state plans. In other states these regulations do not apply to public agencies. OSHA has regulations for response to hazardous materials incidents, confined space rescue, respiratory protection, and fire brigades. Within the Occupational Safety and Health Act is the General Duty Clause, which has implications for operations not explicitly covered by other OSHA regulations.

Consensus standards provide a guide for safety and health programs. The NFPA publishes a number of specific standards relating to firefighter safety and health. These safety standards are found in the *NFPA 1500* series. A number of other NFPA standards are referenced within the *NFPA 1500* series and therefore become part of them by reference. Although a standard might not be formally adopted, there can be standard of care implications.

A number of other regulations and standards also exist that apply to safety and health, including guidelines from the Centers for Disease Control and Prevention, the National Institute for Occupational Safety and Health, ASTM International, American National Standards Institute standards, the Environmental Protection Agency regulations, and the Ryan White Act. In order to develop a comprehensive safety and health program, these regulations and standards must be consulted. If they are law, they must be complied with; but if not law, they can provide a framework for the safety program that is accepted and supported by national standards and trends.

Concluding Thought: Safety and health are ongoing considerations for which there are many regulations that must be complied with, and many standards that can provide an accepted framework of best practices.

REVIEW QUESTIONS

1. The fire chief has asked you to sit on a committee to update standard operating procedures (SOPs). Consider each of the following SOP subjects and list what regulations or standards you would refer to as you formulated the updated SOPs.
   a. Confined space rescue
   b. Hazardous materials response and operations
   c. Infection control
   d. Incident management system
   e. Self-contained breathing apparatus maintenance and usage
   f. Operations at single-family house fires
2. Discuss the meaning of *standard of care*.

3. Which of the following states do not have state OSHA plans?
    a. Florida
    b. Maryland
    c. California
    d. Kentucky

4. Compare regulations with standards.

5. What OSHA regulation requires two in/two out at a structural fire?
    a. 1910.1030
    b. 1910.134
    c. 1910.156
    d. 1910.120

6. The guidelines published by the Centers for Disease Control and Prevention would help the safety program manager in complying with which OSHA regulation?
    a. 1910.120
    b. 1910.156
    c. 1910.134
    d. 1910.1030

7. In effect, 40 CFR 311 mirrors OSHA 1910.120.
    a. True
    b. False

8. NIOSH is part of the Centers for Disease Control and Prevention and was created as part of the act that created OSHA.
    a. True
    b. False

9. *NFPA 1500* is intended to apply to *all* fire departments including industry fire brigades.
    a. True
    b. False

10. You are the lead instructor in a training class that is going to conduct a live burn. Which of the following NFPA standards would be helpful in planning a safe drill?
    a. *1500*
    b. *1403*
    c. *1410*
    d. all of the above

ACTIVITIES

1. Review your department's purchasing process in terms of equipment and apparatus. Does your department follow NFPA standards?

2. Review your department's procedures for response to confined space and hazardous materials incidents. Are your procedures in compliance with those in the respective regulations?

3. Review *NFPA 1500*, then study your department. Determine where your department stands in terms of compliance.

CHAPTER 3

Risk Management

LEARNING OBJECTIVES

Upon completion of this chapter, you should be able to:

- Define Risk.
- Define Risk Management.
- Discuss the meaning of and the process for risk identification.
- Discuss the meaning of and the process for risk evaluation.
- Discuss the meaning of and the process for risk control.
- Discuss the use of a cause and effect diagram.
- Explain the interrupting-the-accident sequence process.
- Discuss the components and use of the Haddon Matrix.
- Differentiate between passive and active, and voluntary and mandatory risk reduction strategies.
- Apply risk control strategies to an injury problem.

On March 13, 2010, at approximately 0615 hours, a fire fighter drowned after being thrown from a boat when it crashed into a bridge and capsized. The fire fighter was part of a swiftwater rescue team which included his captain and his chief. The chief was operating the boat as the crew attempted to make a fourth trip upstream to rescue civilians who were trapped by fast-rising flood waters. The crew was navigating up an eddy created from the water flooding a street when the motor struck something under the water and diverted them out into the main current. The chief could not steer the motor, nor did it react when he put it into reverse. The boat slammed into a concrete bridge. The back of the boat was immediately pushed down into the water and the bow of the boat flipped up and over the stern throwing all three fire fighters into the frigid water. Two of the fire fighters were able to escape the hydraulics created from the flood water rushing against the arched concrete stanchion on the side of the bridge. For personal protective equipment, the deceased firefighter was wearing a 3-mm neoprene wetsuit which was a sleeveless jumpsuit with a 2-mm neoprene full front zippered wetsuit jacket. He was wearing a sleeveless cotton shirt underneath the jacket and a Type-V rescue personal flotation device and a red protective helmet. His boots were a wetshoe workboot constructed of a 5-mm neoprene upper with a 7-mm neoprene insole. The gloves he was wearing were made of 2-mm neoprene. The personal protective gear ensemble of the other rescuers was identical except they were each wearing a dry suit. All ensembles, including the victim, met the criteria of *NFPA 1670.*

The deceased firefighter was recovered six days later approximately 4 miles from where the boat had capsized.

Key recommendations after analysis of the incident included:

1. The authority having jurisdiction should ensure that fire fighters wear the appropriate PPE for the environment encountered.

2. Authorities having jurisdiction should conduct a hazard and risk assessment analysis of their response areas and incident scene to determine what resources are needed to conduct technical search and rescue at swift water incidents.

A risk management process that begins with local risk identification is a required part of a safety and health program.

Source: Information for this case study was taken in part from CDC/NIOSH Fire Fighter Fatality Investigation Report F2010-09. http://www.cdc.gov/niosh/fire/reports/face201009.html

INTRODUCTION

Understanding risk, risk management, and implementation of a risk management plan are the basic building blocks of the occupational safety and health program. Risk management is recommended in the following Firefighter Life Safety Initiative:

> 3. *Focus greater attention on the integration of risk management with incident management at all levels, including strategic, tactical, and planning responsibilities.*

Risk is defined in *NFPA 1500* as a measure of the probability and severity of adverse effects that result from an exposure to a hazard.[1] After the safety and health manager has reviewed the applicable standards, risk management must be undertaken. Risk management, required as part of *NFPA 1500*, is defined as the process of planning, organizing, directing, and controlling the resources and activities of an organization in order to minimize detrimental effects on that organization.[2]

NOTE Risk Management includes five components: risk identification, risk evaluation, establishment of priorities for action, risk control, and risk management monitoring.

The first four of these concepts are presented in this chapter along with strategies to undertake to complete the process. The fifth component, risk management monitoring, is described in Chapter 11, Safety Program Evaluation.

RISK IDENTIFICATION

Risks simply put could be anything bad that could happen to an organization. A risk might be associated with back injuries at a training exercise, a needlestick exposure at an emergency medical incident, a respiratory injury as a result of an SCBA failure at a fire, or simply a fall in the station. Identifying risks is complex, because in some cases you are trying to predict what could happen. Identification should include information from a number of sources including local and national resources.

risks The resultant outcome of exposure to a hazard.

Local Experience

The local experience can be reviewed in terms of historical incident data and data to help project future risks. Historical data can provide valuable information regarding risk and exposure based on the local history. For example, in analyzing the risk for firefighter injury from falls, the safety manager might review the equipment and procedures used to access aboveground locations. The severity and frequency of the risk is presented in the next section, Risk Evaluation.

NOTE Another resource from local experience in the evaluation of the risk is to question the responders.

Responders usually have the best knowledge of the area and have often identified certain potential risks within their response areas; holding brainstorming or Hazard ID sessions will often identify risk that might otherwise be overlooked. Examples may be a simple risk, such as identifying an area of a highway that is prone to flood during rain storms, or a problem with responding units meeting at certain intersections during a response. Other examples may also include certain characteristics of a building or group of buildings that would present increased risk to responders. Any of these cases are clearly a risk in terms of emergency operations.

Historical and current data are useful for determining what has happened or what could happen today. Of greater difficulty is trying to predict what may happen in the future. Identification of future risk can be facilitated by contacting the local planning department, which can provide information on future road expansion, future industrial/commercial occupancies, and other changes that may occur in the community, such as new and remodeled buildings (see **Figure 3-1**). This information will not only help to

FIGURE 3-1

Changes in construction and occupancy occurring in the community should be evaluated as part of risk identification.

© Cengage Learning 2013

identify future risks, but also provide time to plan for the risk. An example might be a town where the tallest building is one story. The town's fire department does an excellent job of risk identification and has developed a risk management plan for the risks that have been identified. However, the city is planning to build a multistory senior citizen residential building. This planned change will significantly change the department's risk management plan because now the risks include those that are common in high-rise structures.

Identification of Trends

Maintaining currency with emerging trends can also be useful in risk identification. By reading trade journals, searching the Internet, and attending conferences and seminars, the safety manager can see emerging trends. For example, the trend for fire departments to become involved in emergency medical response clearly created additional risks; the move toward hazardous material response and control created additional risks. The trend of building houses and apartments with lightweight truss roof construction increased the risks during interior firefighting. The move toward placing supplemental restraints (airbags) in cars increases the risk to responders to vehicle accidents. The threat of terror-related attacks has resulted in additional risks to emergency responders.

Trend analysis can help to identify future as well as current risk. The trend of reducing staffing and increasing services provided by emergency service organizations is certain to create additional risks. Some EMS systems now provide inoculations. Should your department include this involvement in the future, getting stuck with a needle and the concern for cross contamination may be an identified future risk.

Safety Audit

The safety audit can be another way of identifying risk within an organization. The audit should be based on some type of checklist in which a reviewer goes through the organization and checks for compliance. Commonly in emergency services safety programs, the *NFPA 1500* standard is used as a benchmark. Deficiencies are noted and any associated risks identified. For example, the *NFPA 1500* standard requires that all newly purchased protective gloves comply with the current NFPA standard covering fire service gloves (*NFPA 1971*). If the safety audit reveals that half of the department's gloves have not been purchased under the new standard, the changes in the new standard should be reviewed against the quality of the existing gloves and the potential risk of hand injury should be identified. Safety audits can be performed in-house by safety staff or contracted to an outside safety consultant.

Reviewing Previous Injury Experience

NOTE Clearly one of the best ways to identify risk in terms of safety and health is to review past statistics.

Within an organization this review can come from injury reports, workers' compensation history, or, if required, the OSHA illness/injury logs. As described in Chapter 1, it is also important to compare the local statistics to those nationwide and to form an understanding of why they may differ. The review of previous injuries complements the risk identification process basically because, if it happened and it caused an injury, it must be a risk. The one downfall in the process is that you are looking at a history and just because something has happened, it may never happen again—or just because something has not happened, does not mean it will not happen in the future.

Once risks are identified through the four processes described previously, it is useful to categorize the risks. Categories might include responding, in station, training, fire or EMS emergency, natural or man-made disasters, and hazardous materials incidents.

NOTE Categories help the safety manager to later sort the risks.

RISK EVALUATION

Once the risks are identified, they should be recorded for future reference because the next step in the process is to evaluate the risks. For risk evaluation, we introduce two risk management terms, *frequency* and *severity*. Risk evaluation is done from these two perspectives.

Frequency

Frequency is how often a particular risk is likely to occur, be it a fall or an auto collision.

NOTE The safety manager must evaluate how often a risk will occur in order to better assign the risk a priority later.

Unfortunately, there is no hard-and-fast rule for what an acceptable frequency is. It is a somewhat subjective measure that is very dependent on local conditions and the person making the evaluation. For this reason, numerical measures are not assigned to frequency, but instead just a high, medium, or low rating.

For example, suppose our department, on any given day, has 100 emergency vehicles on the highways responding to incidents. Using historical data we determine that there are 50 traffic accidents per year (about one a week), involving the emergency

frequency How often a risk occurs or is expected to occur.

vehicles. Is this frequency high, medium, or low? What about a department with 25 emergency vehicles and the same accident rate? Sounds high? So maybe in the smaller department the risk would get a high frequency rating. However, what if to prevent accidents the first department had just spent a year training drivers in modified responses to nonemergency alarms, installed intersection control lights, and adopted a policy requiring a complete stop at all controlled intersections? This 50 then would be considered a high frequency for them as well.

One way to determine if a frequency is high, medium, or low is to benchmark the local experience to that nationwide or to that of similar-size organizations. For example, if the safety manager for the smaller department selected similar-sized departments with similar demographics and a similar number of vehicles on the road and found out the average number of accidents per year was 100, then the safety manager may rate his department's frequency as low.

Severity

Severity is a measurement of how great the loss or the consequences of the loss will be. There are two variables in the measurement of severity, but again some is in the judgment of the person measuring. As with frequency, a measurement of high, medium, and low is used as opposed to a numerical rating. The following factors must also be considered when determining the severity element:

severity How severe the result is when a risk occurs.

Costs

Costs of a risk are measurable; however, cost can be both direct and indirect. Direct costs might include the costs of medical treatment, the overtime paid to cover a vacancy on a crew, or the cost of replacing equipment. Indirect costs are more difficult to measure, such as the loss of productivity, the loss of using the equipment, stress-related concerns of coworkers, morale, and possibly the cost of replacing the employee.

Organizational Impact

How would the risk impact the organization? For example, the collision of a department's only ambulance would put that department out of the EMS transport business until a replacement could be put into service. Smaller departments could be crippled by an incident in which multiple employees are injured. Also included in organizational impact is the time it would take to recover, which might be the time it takes to rehire or retrain personnel or replace an emergency vehicle.

Establishment of Priorities for Action

When the program manager combines frequency and severity, the result can help to determine priorities to which risk control measures can be applied. It is common and

TABLE 3-1 Frequency and Severity of Identified Risks

RISK	FREQUENCY	SEVERITY
Collisions while responding	L	H
Trapped during firefighting	L	H
Sprains and strains associated with lifting during medical responses	H	M

© Cengage Learning 2013

necessary to prioritize the risks, as seldom are time, money, and other required resources sufficient to control all of them.

NOTE In prioritizing, it is important to remember that these two measures are somewhat subjective and very organizationally specific. However they can be quite useful.

Let us return to our example of the town that is about to get a high-rise building for senior citizens. For simplicity, let us say that we have determined three risks associated with this change. The risks identified are: collisions while responding, trapped during firefighting, and sprains and strains associated with lifting during medical responses. A quick table can be put together (see **Table 3-1**) and an analysis made.

These ratings may have been given after significant study. For example, the building is to be built across from the EMS station; therefore the risk of a vehicle accident en route would be low. However, if an accident were to occur, the severity would be high because the department only has one ambulance. The building is being built to current fire codes and has a state-of-the-art private fire protection and detection system. Although the severity of becoming trapped in a high-rise is high, the potential for this to occur in this building is low. Finally, the safety manager in this case called a fire/EMS department in another town that has a similar building. That department reported that it responds to a minimum of one EMS incident per day at their senior high-rise and usually transports someone. The safety manager further researched his department statistics and found that the department had a higher-than-average back injury problem. Therefore, in the high-rise case, the priority risk should be the prevention of back injuries. This is not to say that the other risks are not important; in fact, risks in every case with a high severity risk, regardless of frequency, should be given consideration for high priority for action.

RISK CONTROL

Once the risk identification and evaluation process has been completed, the safety manager must develop ways to control the risk. **Risk control** can be divided into three broad categories: risk avoidance, risk reduction, and risk transfer.

risk control A common approach to risk management in which measures and processes are implemented to help control the number and the severity of losses or consequences of risk to the organization.

FIGURE 3-2
Sharps boxes are used to avoid needle-stick injuries.

Risk Avoidance

To avoid risk is simply not to do the task with which the risk is associated, but this option is not usually available to the emergency response profession. By the nature of the profession, we are called when things have already gone wrong. Many times it is not possible to avoid the risk. For example, the risk of steam burns from fighting an interior fire could be completely eliminated if the fire department chose not to enter burning dwellings. If the risk of exposure to infectious diseases were a priority, we could avoid the risk by not responding to calls involving blood products. But it would be difficult to justify our existence if we chose this avoidance route and did not respond to these types of incidents.

However, some risks can be dealt with successfully by avoiding the risk. For example, a risk associated with starting an intravenous line (IV) is getting stuck with the needle when trying to resheath it. So to avoid the risk, we develop a policy that sets the procedure for dirty needles to be put directly into a sharps box without resheathing them and using needleless systems when possible (see **Figure 3-2**). A more costly endeavor might be the risk associated with backing the fire apparatus into the bays. To avoid the risk, the department could design future stations with drive-through bays.

Risk Reduction

If the risk cannot be avoided, then the next best effort that can be put forth is to reduce the risk or minimize the chance that it will occur. There are many ways to develop risk reduction measures, and there may be several ways to control a particular risk.

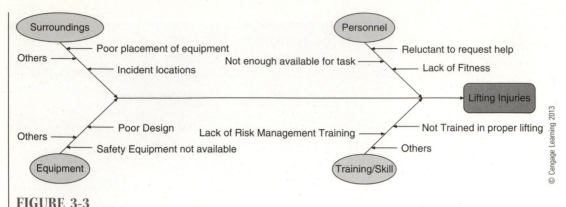

FIGURE 3-3

An example fishbone diagram to analyze the causes of lifting injuries.

NOTE Any development of risk reduction measures must involve input from the employees who are affected.

Risk reduction measures that are not or cannot be followed are useless.

Cause-and-effect diagrams are a useful tool in a risk-reduction process (see **Figure 3-3**). Known as Ishikawa diagrams, they are sometimes called a fishbone diagram. The risk to be reduced is written in a box at the right-hand end of a long arrow. The possible causes of that effect are then listed and connected to the effect like bones connected to the backbone of a fish. It is then possible to look at all of the factors relating to that cause. The level of detail required will vary according to the complexity of the risk. However, the more detailed and specific you can be on the cause, the better the result.

cause-and-effect diagrams Are causal diagrams that show the causes of a certain event.

Interrupting the Accident Sequence

One method of developing risk reduction is to view the injury process as a series of interruptible events. The accident sequence can be described in terms of five factors: the social environment, human factors, unsafe acts or conditions, accident, and the injury.

NOTE The principle behind this thinking is that any interruption in one of the factors will prevent the injury or risk from occurring.

Heinrich[3] makes this theory analogous with dominos standing on end. If one domino is pushed, they all will fall. But if the center domino is removed, the falling will stop there.

In using this concept to generate risk reduction measures, the safety manager can look at past and predicted future risks and break the sequence down into events. Then

TABLE 3-2 Classification of Events Leading to Back Injury

Event	Social Environment	Human Factor	Unsafe Act or Condition	Accident	Injury
Location of equipment on apparatus			X		
Fitness of the firefighter		X			
Not wanting to ask for help	X				
Muscle strain					X
Overexertion				X	

© Cengage Learning 2013

these events can be classified according to the five factors in the sequence and control measures developed.

Let us take, for example, back injuries that occur from lifting heavy equipment from overhead compartments on the apparatus. The safety committee breaks down the events that led up to the injury, and then each event is classified. **Table 3-2**, which is illustrative and not exhaustive, shows the result of this process.

Once the group has brainstormed all events and classified them, control measures can be developed. In this case, procedures could be written that require two persons to lift heavy equipment, apparatus can be designed for heavy equipment to be stored low, and training could be undertaken to improve back strength and develop proper lifting techniques. Although this example is simple, the process can be used on much more complex problems.

The Haddon Matrix

Another method of developing risk reduction strategies can be borrowed from a long-time injury prevention expert and developer of the **Haddon matrix**, Dr. William Haddon. Haddon based the **matrix** on the process of analyzing injuries. The matrix implies that there are four components of the injury process and three time intervals. The four components are the human factors, the energy vector, the physical environment, and the social environment.

NOTE Human factors are related to the employee or person involved in the injury. Human factors can contribute to the injury from three perspectives: poor attitude, inadequate training and education, or lack of fitness for the task.

Haddon matrix A 4 × 3 matrix used to help analyze injuries and accidents in an attempt to determine processes designed to reduce them.

matrix A chart used to categorize actions or events for analysis.

The energy vector is what causes the injury. For a burn it would be the heat of fire or steam; for a back strain it might be the patient or piece of equipment, and for a laceration it could be the glass from a window or jagged metal on a damaged automobile. The physical environment is where the risk is likely to occur. This factor includes weather conditions, seating area on fire apparatus, a natural disaster, or placement of equipment on the fire apparatus. The fourth component is the social environment, which involves the organization's climate, the values and beliefs of employees, and peer pressure.

The preevent time interval is studied to try to prevent the injury or risk from occurring in the first place. The event interval focuses on the risk at the time it is occurring; the strategies are aimed at reducing the exposure or minimizing the risk. The postevent interval is after the event has occurred and focuses on early notice that an event has occurred, proper care if necessary, and proper rehabilitation and recovery.

Risk reduction measures can be generated using this matrix. Although some of the control measures that could be generated may not be practical, many will be. A good way to use this tool is to select a priority risk and have a group, often the safety committee, brainstorm control methods that would fit into each of the 12 blocks in the matrix. The committee or group can then go back and eliminate those that are not practical and further research those that are.

A result of this process is shown in **Table 3-3**, which depicts a partially completed matrix attempting to minimize burn injuries after a brainstorming session by department officers and the safety program manager. This table is illustrative, not exhaustive. Local conditions and procedures may not allow for many of the measures to be implemented.

The Haddon thinking goes yet another step in the risk reduction strategy process. Once the measures that are practical are selected, the measures are further divided as to whether they are passive or active, and voluntary or mandatory. Passive here means that you do not have to do anything special for the control measure to occur; active means that you must physically do something. Voluntary measures allow people to choose, whereas mandatory measures are mandated by laws, rules, or regulations. It is evident that the most effective measures are those that are mandatory yet passive. Using PASS devices is common throughout the fire service. When they first came out, some individuals purchased them for themselves. This risk reduction method would have been considered voluntary and active. It is hard to determine how effective it was. Now PASS devices are a mandatory part of safety standards. However, until recently they required an active process, namely turning them on. Again, effectiveness was in question, because many were not activated while fighting a fire. Recent improvements have the PASS device turn on with the breathing apparatus, hence a passive process and an increase in effectiveness.

Some of the risk reduction measures from the burn example in **Table 3-3** are further categorized in **Table 3-4**. This table is also for illustrative purposes and is not to be

TABLE 3-3 Example of a Completed Matrix

	HUMAN	ENERGY	PHYSICAL ENVIRONMENT	SOCIAL ENVIRONMENT
Preevent	Train firefighters and company officers to recognize flashover conditions Have incident commanders make a risk analysis before committing to interior attacks on buildings not worth saving	Do not go in	Separate from cause Use proper hose stream and adequate gallons per minute Follow good tactics, including ventilation	Education of public to be safer with fire Less peer pressure about wearing personal protective equipment (PPE)
Event	Train on survival techniques and rapid escape Have a rapid intervention team at the ready Wear a self-on PASS device	Wear full PPE	Have an escape route	Less aggressive attacks
Postevent	Training in burn care	Make others aware of fireground or interior conditions	EMS at scene for rapid care and transport	Proper follow-up care and rehabilitation

© Cengage Learning 2013

TABLE 3-4 A Completed Matrix Comparing Voluntary and Mandatory Passive and Active Measures

	VOLUNTARY	MANDATORY
Passive	EMS at scene for rapid care and transport	Wear a self-on PASS device
Active	Use proper hose stream and adequate gallons per minute Have an escape route	Wear full PPE Use proper hose stream and adequate gallons per minute

© Cengage Learning 2013

considered exhaustive. Note that some control measures may fit into more than one category.

Using the 4 × 3 and 2 × 2 matrices can be a valuable tool to assist the safety program manager in developing risk control methods. It is easy to understand and provides the categorization necessary to formulate effective strategies.

Risk Transfer

Risk transfer in emergency services is most commonly associated with the purchase of insurance policies. The transfer of risk does nothing to prevent the risk and only limits the amount of financial exposure for the organization. The transfer of risk would do very little in terms of the time it takes to hire a replacement firefighter for one who was injured and retired on a disability pension. Risk transfer is closely related to the safety and health program, as the amount that is paid for insurance is based on previous experience, including injuries, accidents, and other risks.

Risk transfer through insurance is common. Insurance carriers are interested in the operation of the department from a risk management standpoint, not only to reduce claims, but also because in today's competitive environment many organizations bid their insurance coverage. The insurance company, therefore, also has an interest in keeping premiums low. Expect an insurance company to look favorably on initiatives such as driver training programs, SOPs, drug-free workplace policies, and safety committees, to name a few.

However, the issue of risk transfer through insurance is a complex one, well beyond the scope of this book, so only a brief explanation of different types of insurance is provided.

> **risk transfer** The process of letting someone else assume the risk; for example, buying auto insurance transfers the consequences of an accident to the insurance company.

Workers' Compensation Insurance

Workers' compensation insurance is the insurance that covers employees injured while on the job. Each state has different requirements for workers' compensation, so there is a great deal of variation in requirements and benefits. Generally, workers' compensation provides for the payment of medical expenses required after an illness or injury suffered on the job. Workers' compensation also pays a wage to the employee to cover time off following a job-related illness or injury. Because the wage that workers' compensation provides is only a portion of the employee's normal salary, there is incentive to get back to work. Some employment agreements provide for 100% of salary for a specified period of time after an injury or illness suffered on the job. In this case, the employer is required to make up the difference and may carry a secondary insurance policy.

Management Liability

Sometimes referred to as errors and omissions, management liability insurance is designed to cover the actions, or lack thereof, of employees while performing duties on behalf of the employer.

General Liability

General liability insurance is designed to protect the organization from a property loss. Although coverage varies, protection against theft, fire, storms, and such are usually provided.

Vehicle Insurance

Vehicle insurance covers the organization's vehicles from damage or theft. Vehicle insurance policies also protect the insured from liability caused by the vehicle or its operator. Vehicle insurance generally does not cover an on-duty employee from an injury standpoint. Required medical treatment after a collision is covered by workers' compensation.

Regardless of an organization's safety record or safety and health plan, risk transfer is necessary. The cost of a single incident, such as a vehicle collision, could easily bankrupt a department without insurance. If injuries occurred, there would be property loss, medical expenses and lost time from injured employees, and liability for injuries to occupants of the other vehicles involved.

SUMMARY

Risk management is a five-step process consisting of risk identification, risk evaluation, establishment of priorities for action, risk control, and risk management monitoring. Risk identification, the first step, allows the safety manager both to identify risks from a historical basis and to predict those that are likely to occur in the future. Risk identification is done by analyzing local experience, identifying trends, conducting a safety audit, and reviewing the previous injury experience.

Risk evaluation is a measure of both frequency (how often a risk could occur) and severity (how bad it will be if it does). Combining these perspectives together, the evaluation of risk helps the safety manager establish action priorities and risk control programs.

Risk control is the third step in the risk management process. Risk control can be accomplished by avoiding the risk, reducing the risk, or transferring the risk. In emergency services, risk avoidance has narrow application but in some cases can be useful. Risk reduction, however, can be used aggressively. The tools used to develop risk reduction strategies are cause and effect diagrams, interruption of the accident sequence, and the Haddon matrix. Cause-and-effect or fishbone diagrams help get to the root causes of a risk. Interruption of the accident sequence requires the classification of the accident events in order to develop measures to interrupt the process at some point. The Haddon matrix is a 4 × 3 matrix that considers four factors of three time intervals. The Haddon process further classifies the prevention measures into a 2 × 2 matrix as being active or passive and voluntary or mandatory.

Risk transfer is the process of transferring the risk to someone else. For the purpose of this text, the transfer would be to an insurance company. There are several types of insurance such as workers' compensation, vehicle damage and liability, property damage and loss, and errors and omission. Requirements for insurance are varied and beyond the scope of this text. Readers are encouraged to check local and state laws for insurance requirements.

Concluding Thought: Safety and health program management is not only subjective. It has objective components that are the foundation for local decision making.

REVIEW QUESTIONS

1. List, in order, the five steps in the risk management process required in *NFPA 1500*.
2. What four areas can be examined when performing the risk identification process?
3. Explain the difference in the terms *frequency* and *severity*.
4. Give an example of risk avoidance.
5. List and explain, in general terms, the four kinds of insurance coverage.
6. Explain the value of using a fishbone diagram in risk determining risk control strategies.
7. What are the 12 components of the Haddon matrix?
8. Which of the following would be most effective as a risk reduction measure?
   a. Voluntary/Active
   b. Mandatory/Passive
   c. Mandatory/Active
   d. Voluntary/Passive
9. Give two examples of human factors in the Haddon matrix.
10. Define the social environment.

ACTIVITIES

1. Select an injury class from your department's statistics, for example, back injuries while advancing hose lines. Apply the Haddon matrix and develop some risk control measures.
2. Using the same risk as in Activity 1, apply the interrupting-the-accident sequence process. Compare the similarities and differences. Which process produced better results?
3. Using the same risk as in Activity 1, brainstorm with coworkers or classmates to develop a cause and effect diagram.

NOTES

1. *NFPA 1500, Standard of Fire Department Occupational Safety and Health Program*, 2007.
2. *NFPA 1500, Standard of Fire Department Occupational Safety and Health Program*, 2007.
3. Heinrich, H. W., *Industrial Accident Prevention: A Scientific Approach* (4th ed.), New York: McGraw-Hill, 1959.

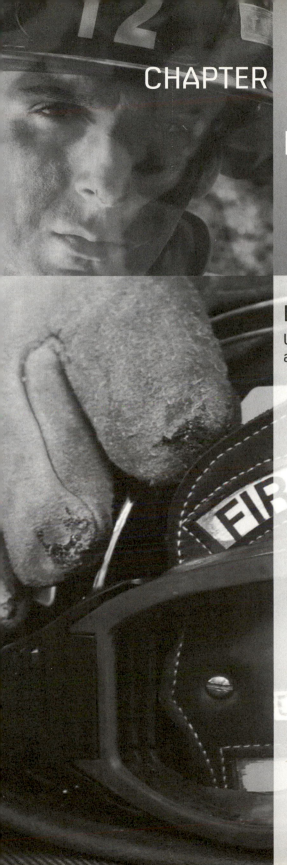

CHAPTER 4

Preincident Safety

LEARNING OBJECTIVES

Upon completion of this chapter, you should be able to:

- Describe safety considerations in the emergency response station.
- Explain safety considerations as they apply to the emergency response vehicle.
- List the components of an effective response safety plan.
- Describe the components of a preincident planning process.
- List the information that should be provided by the preincident plan.
- Describe the considerations for safety while training.
- Define the components of a wellness/fitness plan.
- Describe the considerations for interagency coordination as it applies to health and safety.

On January 9, 2009, a lieutenant died and three fire fighters were injured, when the ladder truck they were riding in failed to stop while traveling down a hill. The ladder crew had just cleared from a medical assist call prior to the incident. The driver turned left out of a parking lot and immediately had to stop and back up to allow the aerial ladder overhanging the cab enough clearance to pass a utility pole on the street. After clearing the pole, the driver immediately negotiated a right downward curve. He then applied the brake pedal and had no response. The driver told the victim he had "no brakes" as the ladder truck gained speed descending the hill. The driver tried placing the ladder truck in neutral and applying the parking (maxi) brake to slow the apparatus down, but there was no response. The driver was able to navigate the ladder truck through a busy intersection, crashing through two parked cars and a brick wall, before coming to rest within a multi-story residential complex. The victim was pronounced dead at the scene; the driver was extricated and transported to a local hospital where he was treated for serious traumatic injuries. The two fire fighters riding in the crew compartment were also transported to the hospital with minor injuries. After analysis, the NIOSH investigators recommended the following:

1. Ensure that preventive maintenance programs are developed and implemented for fire apparatus according to manufacturer guidelines/specifications and national consensus standards.

2. Ensure that preventive maintenance on fire apparatus is performed and/or overseen by qualified personnel who meet the certification requirements outlined in *NFPA 1071, Standard for Emergency Vehicle Technician Professional Qualifications.*

3. Develop and utilize policies and procedures that monitor preventive maintenance and other automotive services performed by vendors.

4. Work with local, state, and federal authorities having jurisdiction (AHJ) in adopting an air brake endorsement for noncommercial driver's licenses that would require individuals operating air brake–equipped fire apparatus to verify their knowledge and skill proficiency (e.g., air brake system, air pressure gauges and alarms, and pre-trip inspection) prior to operating the fire apparatus.

5. Ensure that current driver training programs provide ample classroom instruction, behind-the-wheel driving, procedures for driver clearance, and annual refresher training.

6. Develop and implement fire apparatus inspection procedures and check sheets for their fleet, provide a systematic approach for communicating and receiving inspection sheets from the field, and institute a system to file and track fleet records (e.g., inspection sheets, work orders, repairs, apparatus specifications, and maintenance).

7. Ensure that policies and procedures are developed and enforced on the use of seat belts.

8. Develop policies and procedures that assist with determining specifications for new apparatus and replacement cycles of existing fire apparatus.

As reflected in this case study, the design, ongoing operation, and preventative maintenance of the emergency response vehicle is essential to provide for the safety and health of the responders.

Source: Information for this case study was taken in-part from CDC/NIOSH Fire Fighter Fatality Investigation Report *F2009-05.* http://www.cdc.gov/niosh/fire/reports/face200905.html

INTRODUCTION

During the risk identification process, a number of risks can be identified that occur before an incident does. These risks focus on issues relating to the workplace, the equipment or vehicle being used, and the human resource that is using it or working in the workplace. In this chapter, the focus is on risks that occur prior to an incident or at the time of an incident. Many of these risks are the easiest to manage because, with few exceptions, the risk occurs under a controlled situation or environment.

Preincident health and safety can be divided into seven categories:

1. Station safety
2. Apparatus safety
3. Response safety
4. Preincident planning
5. Safety while training
6. Fitness/wellness
7. Interagency cooperation in safety-related matters

Five of the 16 Firefighter Life Safety Initiatives focus on the subjects that are included in this chapter. They include the following:

6. *Develop and implement national medical and physical fitness standards that are equally applicable to all firefighters, based on the duties they are expected to perform.*

11. *National standards for emergency response policies and procedures should be developed and championed.*

12. *National protocols for response to violent incidents should be developed and championed.*

13. *Firefighters and their families must have access to counseling and psychological support.*

16. *Safety must be a primary consideration in the design of apparatus and equipment.*

STATION CONSIDERATIONS

Consideration for safety at the station is grouped into two broad categories: design and ongoing operations. The design category deals with fire station design prior to construction or at the time of renovation, for example, provisions for an internal fire protection or exhaust emission system (see **Figure 4-1**). Ongoing operations looks at safety and health issues from a day-to-day operational basis, such as wiping up wet spots to prevent falls or station safety inspections.

Design

When it comes to facility safety, public safety agencies should serve as models in the community. Therefore, during the design phase of the fire/EMS station, safety must be an important design consideration. The station must be built to comply with all

FIGURE 4-1
There are many different designs for fire stations. Each should consider responder safety during day-to-day operations.

Courtesy of Palm Harbor Fire Rescue

FIGURE 4-2
Fire stations should be
designed to meet all local
building and fire codes.

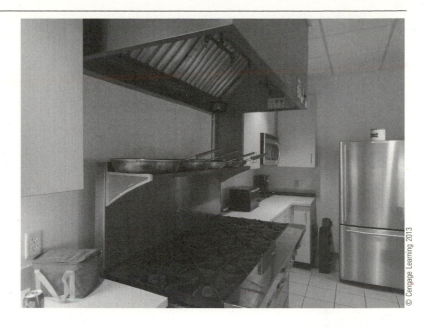

applicable standards and codes. Chapter 9 of *NFPA 1500* requires compliance with all
applicable health, safety, building, and fire codes for fire stations (see **Figure 4-2**). Stations must be designed with fire detection and protection systems such as sprinkler systems, smoke and heat detectors, and carbon monoxide detectors. Emerging trends of
fire station design also include provisions to be more environmentally friendly, or green.
These new green systems will require evaluation on a case-by-case basis for any health
and safety implications.

Once the design is in compliance with local codes and standards, it is important to
look at the physical design. Proper routes of travel must be provided from living areas
to the apparatus bays. These routes of travel must be direct, without obstruction, and
lighted for night operations. Sleeping areas must be designed so that after dark, low-
level lighting is provided to prevent trip and fall injuries. If the station is multistory
and chutes or poles are used, there must be protection from employees falling down
them by accident, and new employees must be trained on the proper use of these devices. The bays should be provided with proper drainage to limit standing water, and
walk areas should be painted using a nonslip paint. Bathroom floor and shower areas
should be nonslip surfaces as well.

Storage areas for protective clothing must allow for ventilation and must not be in
living areas. Unless contracted out, a laundry facility may be provided.

NOTE By OSHA regulation, there must be an area for the cleanup of
materials contaminated with bloodborne pathogens.

The bays should have some form of exhaust emission system. Exhaust systems may be the type that attach to the vehicle's exhaust by a tube and are vented to the outside. A second system vents the entire apparatus bay area, using roof or wall-mounted exhaust fans and louvers in the bay doors. This system is useful in colder climates in which small equipment, such as rescue tools and saws, must be run inside the bays. Yet another exhaust elimination system is not part of the station, but attaches and remains on the emergency vehicle exhaust.

NOTE The station design process should include input from the safety committee and the crews themselves.

A good approach is to have a building committee made of the various building users as well as the department's health and safety officer. The crew will be able to pinpoint safety issues in the design of a building (see **Table 4-1**).

It clearly is much easier and less costly to design the previously mentioned features into a new building prior to construction. However, most of the facilities that the safety manager deals with today have long been constructed. Some are more than 100 years old. The safety manager, in conjunction with fire department management and the safety committee, must prioritize improvements and try to retrofit the station with the necessary safety features over several budget years. Modifications that are required by regulation may have to be made early on.

Sufficient parking should also be provided, particularly for volunteer departments where members respond to the station for the apparatus. Responders should be able to park where they can exit their vehicles and access the station safely. The driveway access should be well lit for any night operations and provide room for apparatus checks. If practical training is to be held at the facility, there must also be a consideration for these needs and associated hazards.

Ongoing Operations

Of additional concern regarding station safety—and probably more important—are the procedures and safety measures relating to the day-to-day operations of the station and the inhabitants, the responders, and the public that might visit. The station is generally a controlled environment, so injuries that occur at the station could often easily have been prevented.

NOTE First and foremost, the fire station should have a complete fire inspection at least annually in accordance with local code.

Fire protection systems should be certified as recommended or required in the code. Smoke, heat, and carbon monoxide detectors should be checked according to manufacturers' recommendations. Generally, the on-duty crew can perform these

TABLE 4-1 Example Station Design Checklist

	YES	NO
Applicable Code Compliance		
Health		
Safety		
Building/life safety		
Fire prevention		
Physical Design		
Routes of travel—direct, marked, free of obstruction, lighted		
Sleeping areas—low-level lighting provided		
Bathroom areas—nonslip surfaces		
Poles or chutes protected		
Apparatus bays—drainage		
Apparatus bays—lighting		
Apparatus bays—exhaust emission system		
Apparatus bays—nonslip surface		
Apparatus bays—doors: Self-reverse safety system		
Apparatus bay access—drive through preferred		
Biohazard cleanup room/area		
PPE storage room with ventilation—not in living areas		
Station access—proper design—ample parking		
Proper and ample storage areas provided		
Safety and Security		
Ability to secure station		
Fire alarm system with smoke and heat detectors		
Sprinkler system		
Carbon monoxide detectors		
Provisions to protect employees and equipment from extreme weather		
Back-up power		
Automatic shutoff for cooking appliances		

© Cengage Learning 2013

inspections, sometimes with assistance from the fire prevention division, and by using standard forms (see **Figure 4-3**). Local OSHA or state departments of labor will usually help a department identify safety problems in the station, as will many insurance carriers, and most will provide facility inspection forms.

Aside from the fire safety concern, station inspections that look for hazards and safety issues should be conducted daily. Part of the shift change process should require

STATION INSPECTION REPORT

Station _____ **Shift** _____ **Date** _____

Captain _____ **Battalion Chief** _____ **Division Chief** _____

✔—O.K.
✗—See Remarks

1. Personnel

Uniform	
Protective Clothing	
Grooming & Cleanliness	
Driver's License	
I.D. Card	
E.C. Card	

5. Safety

Overhead Storage	
Caution Signs	
Safety Bulletin Board	
Tool Storage	
Combustible Storage	
Insecticide Chemical Stor.	
SCBA Tests	
Extinguisher(s)	
Smoke Detector(s)	
RADEF Equipment	

6. Exterior

Paint	
Doors	
Sidewalks	
Ramps	
Parking Area	
Yard	
Hose Tower	
Gas Pump	
Flag Pole	

2. Office

Files	
Maps	
Log Book	
Shift Change Log	
Desk	
Bulletin Board	
Tourist Information	

3. Engine Hose

Floor	
Work Bench	
Hose Rack	
Storage	
Turnouts	
Apparatus	

7. Interior

	Office	Kitchen	Day Room	Dormitory	Rest Room	Heater Room		
General Appearance								
Cleanliness								
Floors								
Windows								
Lights								
Furniture								
Storage Area(s)								
Woodwork								
Walls								
Ceiling								

4. Programs

Hazard Reduction	
In-Service Inspections	
Pre-Fire Plans	
Water Systems	

Remarks _____

© Cengage Learning 2013

FIGURE 4-3
An example of a station inspection record.

the on-coming supervisor to walk the station area and look for safety or health problems. These might include wet spots on floors, the proper operation of an emergency bay door stop, or burnt-out lights in a bedroom. Further department procedures should forbid unsafe acts such as creating tripping hazards with debris or electrical cords. The procedures can also be proactive, such as using "Wet Floor" signs after mopping a floor. It is interesting that on the fireground a firefighter will climb 15 feet to a second-floor window for an operation and, using a leg lock, lock into the ladder. But in the station that same firefighter will climb 25 feet on a stepladder to change a light bulb and not consider tying off. Procedures and training should prohibit this type of behavior. Some examples of station safety policies are provided in **Text Box 4-1** and **Text Box 4-2**.

APPARATUS SAFETY

Safety involving apparatus also has some preincident considerations. Like stations, apparatus considerations can also be categorized into design and ongoing operations. Design concerns are those that are analyzed and controlled at the time of purchase. Ongoing concerns are those that affect day-to-day operations. One major issue associated with apparatus is driving and responding to emergencies; that issue is addressed in the next section.

Design

Emergency service vehicles should be designed to meet all applicable standards and safety should be a primary concern. There are NFPA standards that govern requirements for fire apparatus (see **Table 4-2**), many of which are based on firefighter safety needs. There are also requirements placed on the vehicles by the Department of Transportation, because the vehicle, in most cases, will travel on public roads. Ambulances also have federal regulations from the Department of Transportation (see **Figure 4-4**).

Although designing and meeting the minimum requirements stated in the standards provides some level of safety, the safety manager and safety committee should have direct involvement in the design process. Again, as described in the preceding section, a group of users brought together to form an apparatus design committee will be useful. A risk identification and analysis should be part of the design phase. For example, the placement of equipment affects compartmentalization requirements and location. How will ladders be stored? If attached to the side of the apparatus, will they be too high to be safely removed? How about hose bed height for both supply line and attack lines? Can they be safely deployed and, just as important, safely repacked? Many modern apparatus designs have moved the attack lines to lower points to facilitate safe deployment and repacking.

The standard allows for the first step into the cab to be 24 inches from the ground. Is this low enough for a department that has identified ankle injuries caused by getting out of the cab as a risk? Or could another step be installed that might lower

TEXT BOX 4-1

EXAMPLE: FACILITY SAFETY

Purpose

Fire stations, like all structures, are subject to fire and safety hazards. This standard is designed to keep fire stations as safe as possible to protect the lives and health of the personnel within.

Definition

Facility safety is the result of regular inspections for various hazards and conscientious personnel who are always on the lookout for dangerous situations and acts.

Objectives

- To schedule regular monthly and annual inspections of all fire stations.
- To provide for the immediate correction, or to be corrected as soon as possible, of all hazards and code violations found during said inspections.
- For all personnel to become safety conscious about their individual stations.

Station Inspection

- *Annual inspections.* An annual inspection of each station will be done by a company in-service inspector (minimum) and will be accompanied by the shift commander or his designee. The inspection will be done on the first Thursday of September using the standard Safety Survey Form. Copies of the form will be given to the Fire Prevention Bureau and to the shift commander or his designee. Corrections will be made as soon as possible.

- *Monthly Inspections.* A monthly walk-through inspection will be done on the first Thursday of each month using forms provided. The intent of this inspection is to identify unsafe conditions, check that exit lights and detectors work properly, and that extinguishers and the Ansul System are properly charged. Department fire sprinkler forms shall also be completed and attached to the inspection form. Hazards or violations will be reported in writing to the Fire Prevention Bureau and the shift commander or his designee. Corrections will be made as soon as possible.

TEXT BOX
4-2

EXAMPLE: SAFETY SOP

Office Safety (All Personnel)

A. Do not connect multiple electrical devices to a single outlet.

B. Do not use extension or other power cords that are cut, frayed, or damaged.

C. Close file and desk drawers when unattended. Do not open more than one drawer at a time and close it when done.

D. Put heavy files in bottom drawers of file cabinets to prevent cabinets from tipping over.

E. Do not tilt your chair back on two legs.

F. Do not use chairs, boxes, or other items as improvised climbing devices.

G. Turn off the machine and disconnect electrical power before attempting to adjust or clear electrical office equipment.

H. Do not remove, bypass, or tamper with electrical equipment fuses, switches, or safeguards.

I. Do not place your fingers in or near the feed of a paper shredder. Verify guards are in place and working prior to use.

Station Safety (All Personnel)

A. Mop/clean up oil, hydraulic fluid, water, or grease from apparatus floors and accesses immediately upon detection.

B. Safety goggles or other provided eye protection shall be worn when operating power equipment.

C. Do not run electrical and other cords across doorways and aisles, or between desks, or create trip hazards.

D. Pick up all foreign objects, such as pencils, bolts, and similar objects, from floors to prevent slipping.

E. Clean up all spills immediately, especially wet spots around drink and coffee machines, in bathrooms, kitchen, and hallways.

F. Use stepladders only on a firm level base, opened to the full position with spreaders locked.

G. Do not leave tools in work areas, walkways, or stairways.

H. Do not walk across wet or oily areas. Avoid the hazard by walking around it.

I. Avoid walking across areas that have been freshly mopped.

J. Do not run on stairs or steps or take two at a time.

(continues)

K. Use handrails.

L. Do not block your own view by carrying large objects.

M. Do not jump from trucks, platforms, scaffolds, ladders, roofs, or other elevated places, regardless of height.

N. Do not take shortcuts. Use aisles, walkways, or sidewalks.

O. Do not jump from, to, or between elevated areas.

© Cengage Learning 2013

TABLE 4-2 NFPA Standards Impacting Apparatus Design and Ongoing Operations

STANDARD NUMBER	TITLE
NFPA 1901	Standard for Automotive Fire Apparatus
NFPA 1906	Standard for Wildland Fire Apparatus
NFPA 1925	Standard on Marine Fire-Fighting Vessels
NFPA 1912	Standard for Fire Apparatus Refurbishing
NFPA 1915	Standard for Fire Apparatus Preventive Maintenance Program
NFPA 1071	Standard for Emergency Vehicle Technician Professional Qualifications
NFPA 1581	Standard on Fire Department Infection Control Program
NFPA 1911	Standard for the Inspection, Maintenance, Testing, and Retirement of In-Service Automotive Fire Apparatus
NFPA 1917	Standard for Automotive Ambulances

© Cengage Learning 2013

the step level to 18 inches? What about the design of the patient compartment of the ambulance? Again, the standards provide some level of safety, but the unit should be designed around local needs and conditions (see **Figure** 4-5).

Remember, manufacturers of emergency vehicles are willing to do just about anything possible to design a vehicle to meet the needs of their customer. However, many times modifications cost money and may be deleted in favor of other requirements for the apparatus. The safety manager's responsibility is to perform a thorough risk analysis and support each and every safety consideration.

Ongoing Operational Concerns

Most departments keep emergency vehicles for a number of years. A lot can be said for a vehicle that is designed properly, but most of the safety issues relating to apparatus are of the day-to-day operational nature.

FIGURE 4-4
Ambulance design is governed by the Department of Transportation in the KKK specifications. Here is an example of interior layout.

Courtesy of E-One, Inc., Ocala, FL.

FIGURE 4-5
Apparatus must be designed to meet standards and regulations and also address local needs and conditions.

© Cengage Learning 2013

Many of the ongoing operational safety objectives can be accomplished by having a strong **preventive maintenance (PM) program** meeting the requirements of *NFPA 1915* (see **Figure 4-6**). The PM program should include daily checks of emergency lights,

preventive maintenance (PM) program An ongoing program for maintenance on vehicles; designed to provide routine care, oil changes and the like, as well as catch minor problems before they become major ones.

FIGURE 4-6

A preventive maintenance program is integral to having a safe response vehicle.

fluids, and the operation of vehicle components including brakes, pumps, transmission, and aerial ladders. The daily check should be performed by the apparatus operator assigned that day. The operator should be trained to look for maintenance problems before they become worse or cause an accident. Using the apparatus operator also allows the operator to become familiar with the vehicle and vehicle components.

A more thorough check should be performed weekly, including running small tools, cleaning the apparatus thoroughly, checking tire wear and pressures, and back-flushing pumps. A monthly check may go even further, requiring the performance of such tasks as changing water in the water tanks.

Another component of the PM program is a semiannual trip to the maintenance professional. This trip can provide for scheduled fluid changes and visual checks of brake linings and other critical vehicle systems. In the case of fire apparatus, pumps and ladders must be tested annually as well.

The *NFPA 1915* standard requires the organization to develop and implement a schedule of service and maintenance for fire apparatus based on the manufacturer's recommendations, local conditions, and operating conditions. It also describes the requirements for various systems and components of the fire apparatus. *NFPA 1915* further requires that the organization develop written criteria for when the apparatus is to be taken out of service. It is required that this criteria consider 49 CFR, Part 390, "Federal Motor Carrier Safety Regulations"; any applicable federal, state, and local regulations; applicable nationally recognized standards; manufacturers' recommendations; and guidelines established by the organization or its designated service and maintenance staff. The apparatus cannot be returned to service until after defects and deficiencies have been corrected. An example policy is shown in **Text Box 4-3**.

TEXT BOX 4-3

VEHICLE OUT OF SERVICE GUIDE

Purpose: To keep our vehicles in safe operating condition and to assist personnel in placing vehicles out of service. This SOP categorizes apparatus repair needs utilizing a red, yellow, and green format of all items that require repair based on seriousness of mechanical or physical condition.

Red (Immediate): Any and all items that would cause a vehicle to be removed from service immediately. Prompt notification of the on-duty officer and shift commander.

A. SAFETY: Brakes, broken belts, elimination of a major component such as an alternator, power steering, water pump, transmission, flat or heavily damaged tire, audible air leak at brake chamber, loss of emergency lighting or more than one signaling device, etc.

B. INCAPACITATING: Any problem found that prevents starting or does not allow vehicle to operate properly after start up.

C. EXPANDING: Any problem, large or small, left unattended that might potentially cause serious damage to the vehicle or any of its major operating systems; for example, items such as serious leaks (i.e., excessive engine oil, transmission fluid, hydraulic oil, transfer case, rear axle, wheel bearings, antifreeze, etc.) and alternators not charging to full capacity.

D. ACCIDENT: Any accident in which the vehicle has sustained heavy damage.

Yellow (As Soon As Possible But Not Immediately): Any and all items that would not cause a vehicle to be removed from service. Requires notification of the on-duty officer and shift commander.

A. Any item that is important and needs prompt attention, but does not require the vehicle to be taken out of service immediately. (Nonaudible slow air leak, worn belts, etc.).

B. Any problem, large or small, left unattended or that may cause serious damage to any of the vehicle's components if not repaired before next scheduled service.

C. Warning devices (door ajar alarms) not working properly.

Green (Can Wait, Or In-House Repair): Any item that should be repaired that cannot be repaired by the driver engineer or shift personnel, but can wait for the next scheduled maintenance appointment. These items need to be documented on the back sheet of the check-off sheets.

(continues)

> A. This would be such items as dash lights, torn upholstery, worn tires, marker light covers, leaking pump packing, or any nonessential components on vehicle that will not hinder operation of the vehicle or cause damage to it by not being repaired immediately.

Courtesy of Palm Harbor Fire Rescue

This type of preventive maintenance program can be applied to any size or any type of department. Whether the apparatus is a fire truck or an ambulance, the same concepts apply.

NOTE Lack of preventive maintenance reduces the useful life of a vehicle, increases maintenance costs, and can have a significant negative effect on the safety of those relying on the vehicle.

Standard operating procedures and training can also help in the day-to-day operational safety process. Procedures identifying the minimum number of personnel to perform certain tasks have been successful in reducing injuries. For example, for reloading the hose, the procedure may require four persons to be on hand: two in the hose bed, one on the back step, and one on the ground. Keeping the apparatus running board free from debris or fluids prevents slip and fall injuries. Proper lighting when working around the vehicle can also reduce injuries.

Apparatus safety procedures should also address hearing conservation. The USFA has published a manual to assist emergency service organizations in developing a hearing conservation program.

Apparatus safety must include risk identification and evaluation, and the implementation of control methods that can only be designed at the local level. **Text Box 4-4** outlines some apparatus safety considerations, although some are related to response. Another resource is the **International Association of Emergency Vehicle Technicians (IAEVT)**, which can provide valuable information on apparatus maintenance and safety operating practices. The IAEVT can be found on the Internet at http://www.iaevt.org/

International Association of Emergency Vehicle Technicians (IAEVT) An organization that bestows professional certification in many areas for persons involved in emergency service vehicle maintenance.

TEXT BOX
4-4

EXAMPLE POLICY: APPARATUS SAFETY

Purpose

The purpose of these standards is to establish safety procedures in regard to apparatus practices, thereby reducing the potential of accidental injury or death to fire department personnel and the general public.

Vehicle and Equipment Inspection Procedures

A. In order to ensure vehicle readiness and the presence of essential equipment, all apparatus and equipment shall be regularly inspected and/or tested by personnel at intervals prescribed on the existing Vehicle Inspection Form. Daily vehicle and equipment checks shall be performed as close to 0800 hours as practicable and their timely completion is to be superseded only by emergency calls or by the order of the officer in charge.

B. Any apparatus or equipment deficiencies shall be noted on the departmental Vehicle/Equipment Deficiency Form. The officer in charge shall also be verbally notified of any such deficiencies. The officer in charge shall then take necessary action to effect repairs or remedy the deficiency. The equipment or apparatus shall be placed out of service should the deficiency be deemed serious or hazardous.

C. All vehicles and equipment shall be outside of the bays while operating (weather permitting).

Driver's Responsibility

A. It is the responsibility of the driver of each fire department vehicle to drive safely and prudently at all times. Vehicles shall be operated in compliance with traffic laws pursuant to Florida Statute 316. Emergency response does not absolve the driver of any responsibility to drive with due caution. The driver of the emergency vehicle is responsible for its safe operation at all times. The officer in charge of the vehicle has responsibility for the safety of all operations.

B. It is the driver's responsibility to ensure that all passengers are safely aboard the apparatus and seated prior to moving the vehicle.

C. When responding to an emergency, emergency warning lights must be on and sirens must be sounded to warn drivers of other vehicles.

D. Intersections present the greatest potential danger to emergency vehicles. When approaching and crossing an intersection with

(continues)

the right-of-way, drivers shall not exceed the posted speed limit. When approaching a negative right-of-way (red light, stop sign, or yield), the vehicle shall come to a complete stop. The vehicle may proceed only when the driver can account for all oncoming traffic and all lanes yielding right-of-way.

E. In order to avoid unnecessary emergency response, the following rules shall apply:

1. When the first unit reports on the scene with "nothing show-ing," "investigating," or the equivalent, any additional units shall continue emergency, but shall not exceed the posted speed limit and shall come to a complete stop before pro-ceeding at every negative right-of-way intersection.

2. The first arriving unit will advise additional units to respond nonemergency whenever appropriate.

F. During an emergency response, fire vehicles should avoid pass-ing one another. If passing is necessary, arrangements should be conducted through radio communications.

G. The unique hazards of driving on or adjacent to the fireground require the driver to use extreme caution and to be alert and pre-pared to react to the unexpected.

H. When stopped at the scene of an incident, vehicles should be placed to protect personnel who may be working in the street and warning lights shall be used to make approaching traffic aware of the incident. At night, vehicle-mounted floodlights and any other lighting available shall be used to illuminate the scene.

I. Backing. When the backing of apparatus is unavoidable, the driver shall use a guide to back and shall not proceed to back un-less the guide is clearly visible in the mirror(s) being used for visualization. Illumination of the backup area with floodlights and/or flashlights shall be used in times of darkness or limited visibility. If no guide is available, the driver shall dismount and walk completely around the apparatus before proceeding to back. Audible backup alarms shall be used during any backing procedure.

Seating/Seat Belts

A. Personnel are to ride in a seated position at all times that the apparatus is in motion and only in seats or positions where seat belts are provided. Seat belts are to be utilized at all times when vehicles are in motion. Under no circumstances will personnel be permitted to ride in a standing position or on any sideboards or tailboards of any apparatus.

B. EXCEPTIONS to above:

1. Seat belt use will not be required during actual fire combat from exterior positions on the brush or water trucks during brush fires. For this reason, brush/water truck drivers shall exercise extreme caution with regard to speed, terrain, overhead clearance, etc. If terrain is extremely rough or other conditions exist that would render exterior positioning unsafe, the rider shall dismount from the vehicle and either proceed on foot or take a seated position in the interior of the cab. This exception does not exempt personnel from the required use of seat belts in brush/water apparatus during routine or emergency over-the-road operations or while traveling point-to-point during off-road operations.

2. Fire department personnel riding in ground or air ambulances or any other vehicles shall wear seat belts when possible.

3. It is obvious that seat belts are impractical during repacking of hose in the hose bed of apparatus. This should not be attempted without adequate personnel.

C. When responding to calls requiring the donning of protective clothing, donning shall be completed prior to response since this procedure would be difficult to complete while seat belts are engaged.

D. If a call requiring the donning of protective clothing is received during routine driving of apparatus, the driver shall proceed to a safe location off the roadway where personnel may don protective clothing prior to continuing response. It is not recommended that personnel wait until arrival at the scene to don protective clothing.

E. Donning of air packs while apparatus is in motion shall be done only in apparatus provided with seating that permits donning while seat belts are engaged and proper seating/posture can be maintained. In apparatus where this is not possible, personnel shall don air packs upon arrival at the scene.

F. Safety bars (e.g., Man Savers) shall be in the down position at all times and shall not be disabled or altered in such a manner that would defeat the intended function.

G. A common cause of ankle/knee injuries is forward (front-facing) dismount from apparatus. All personnel, especially when wearing air packs or carrying heavy loads, should back down off of apparatus using hand-hold handles to assist.

Courtesy of Palm Harbor Fire Rescue

RESPONSE SAFETY

The first consideration in response safety is the selection, training, and capabilities of the driver operator.

NOTE Apparatus driver selection should consider the human aspects such as attitude, knowledge, mental fitness, judgment, physical fitness, and past driving record.

NFPA 1002, Standard for Fire Apparatus Driver/Operator Professional Qualifications, outlines the requirements for operators of fire apparatus in both emergency and nonemergency situations. The standard requires that the driver be licensed to operate the vehicles they are required to drive and requires a medical exam to prove fitness to operate an emergency vehicle. The specific requirements and performance objectives for driving and operating various types of apparatus can be found in the standard. The various types of apparatus include those equipped with an attack or fire pump in Chapter 5 of *NFPA 1002*, aerial devices in Chapter 6, tiller operators in Chapter 7, wildland fire apparatus in Chapter 8, aircraft rescue and firefighting apparatus in Chapter 9, and water supply apparatus in Chapter 10. The standard also requires the organization to evaluate these required performance objectives in the selection of drivers.

In order to meet *NFPA 1002*, most departments perform some type of testing prior to driver selection that measures some or all of these characteristics. Further, prior to or after selection, the driver should be trained in emergency vehicle operations in a vehicle similar to the one he or she will have to drive and operate. This training program should address legal issues, physical forces including controlling the vehicle on a variety of road surfaces, safe stopping distances, vehicle maintenance, departmental safety procedures, general safe driving practices, and specific safe driving practices based on the geographic area and environment. This training must have both classroom and practical components. A driving course is a common approach to the practical phase of this training, but provisions must be made for different driving surfaces, including wet roads. A commonly used driving course setup is provided in **Figure 4-7**.

The department's insurance company may also offer driver training resources; some even have complete programs, including an instructor. The emergency vehicle driver must have the capabilities to control the vehicle and react quickly and appropriately to changing situations. Excessive speed, reckless driving, and failing to slow down or stop going through intersections have been the causes of many emergency vehicle accidents. Policies that prohibit operating an emergency vehicle while under the influence of alcohol or drugs, including prescription medications, must also be in place and enforced.

Once the driver is selected and trained, the program must focus on the response. Consideration for procedures that require the use of seats and seat belts is necessary. The habit of riding tailboards or sidesteps is unacceptable and for the most part has stopped. However, standing in the jump seat area still occurs. The supervisor on the

FIGURE 4-7

A typical obstacle course for emergency services driver training.

emergency vehicle should be required to ensure that all occupants are properly seated and belted prior to the response.

The type of response also has safety implications. An emergency response using warning devices is more dangerous than a response in which the vehicle flows normally with traffic. The safety manager, in conjunction with the department management, should review the types of incidents that the department responds to and determine which may not require emergency response. Many other departments have adopted policies for response in a "no lights and siren" mode for certain situations. Others have reduced lights and siren responses on certain types of incidents for all units except the first due units. Medical Priority dispatch has resulted in non lights and sirens responses to many nonemergency medical incidents. **Text Box 4-5** lists some types of incidents that might be considered for a nonemergency response.

NOTE The environment in which the response takes place can also be analyzed and risks identified.

Traffic patterns and common response routes should be studied. Apparatus accidents have occurred between two fire department vehicles responding to the same incident from different locations. This risk can be reduced with better radio communication and preincident procedures for response. Target intersections can also be

TEXT BOX 4-5

TYPES OF INCIDENTS FOR NONEMERGENCY RESPONSE CONSIDERATION

- Automatic alarms
- Sprinkler alarms
- Natural gas leaks
- Wires down
- Calls for manpower
- Nonemergency medical incidents
- Lockouts
- Investigation of drums, barrels, unknown odors, and such
- Carbon monoxide detector alarms
- Rubbish, weeds, and dumpster fires without exposure
- Move-ups to fire stations
- Broken sprinkler or water pipes
- Smoke detectors
- Manual pull stations
- Refrigerator doors
- Plugging details
- Assist the police
- Keys in running vehicles

Even if a response is dispatched as a nonemergency response, the response can be upgraded to emergency if additional information is received by indicating that life is in danger, persons are injured, there is a working fire, and so forth.

identified. A problem intersection where emergency vehicle collisions or near misses have occurred should be avoided. However, this choice is seldom reasonable. The department could look at intersection control. Traffic signal preemption is popular across the country. There are several types of systems available. In general the traffic signal controller gets a signal from the preemption device and changes the traffic signal light to green for the direction that the emergency vehicle is traveling and red in all other directions. The signal from the preemption device can come from an optical emitter on the response vehicle, the vehicle's siren, or, if the intersection is near the station, it can be hardwired to the station. The use of global positioning system (GPS) is another traffic preemption alternative. Mobile data computers in the response vehicle using GPS can allow the officer to see the location of all of the responding vehicles.

Procedures may require that the vehicle come to a complete stop at every intersection or not exceed the speed limit in residential neighborhoods. The environment is dynamic, traffic patterns change with time of day, and weather can change quickly. Road construction should be planned for and response routes evaluated. Areas along the response route that have poor visibility should be identified prior to the response. Railroad crossings, particularly those that are not controlled, should be noted on maps or the drivers should be trained in their locations. In 2002, a volunteer firefighter/ engineer was fatally injured when the tanker truck he was driving was struck by a freight train as he attempted to traverse a private, unguarded railroad crossing. The victim was returning to the station from a live-burn training exercise. NIOSH made the following recommendations in its report of the incident:

- Ensure that drivers of fire department vehicles come to a complete stop at all unguarded railroad grade crossings during emergency response or nonemergency travel.
- Revise or develop SOGs for safely driving emergency vehicles during emergency response and nonemergency travel by addressing hazards firefighters are likely to encounter, such as railroad crossings.
- Enforce policies that require that all firefighters who ride in emergency fire apparatus are belted securely by seat belts.
- Provide firefighter training on railway traffic safety in communities such as this, in which a high density of railway traffic exists.

PREINCIDENT PLANNING

The value of preincident planning cannot be underscored enough, not only from a safety standpoint, but also from an operational standpoint. Having preincident plans available is like a coach having the playbook at a game.

NOTE In the case of a fire, preincident planning allows the fire department to fight the fire before it happens.

From a safety standpoint, preincident planning can be used to identify and evaluate potential risk long before an incident occurs. For example, preincident planning may reveal that an occupancy uses and stores combustible gases, is constructed of lightweight trusses, or has access problems. Having identified hazards, responders can adjust tactics to limit their exposure to potentially unsafe situations.

NFPA 1620, Recommended Practice for Pre-Incident Planning outlines the necessary components for a preincident planning program to be effective. It covers various types of occupancies as well as the varied kinds of information that should be collected. Before implementation of a comprehensive preincident planning program, this document should be reviewed.

To be effective, the preincident planning process should contain the following components:

- The preincident plan should be on a form used department-wide.
- Preincident planning should be done by the responders so that they can become familiar with the building during the preparation of the plan.
- The process should provide for updating the plan at given time intervals. A good goal would be an annual update. Remember, occupancies change and renovations occur. Target hazards should be identified and receive priority in the planning process. Target hazards may include:
  - Health care facilities
  - Large industrial occupancies
  - Facilities using or storing hazardous materials
  - High-rise buildings
  - Malls or other high occupancy locations, such as office complexes or stadiums
  - Schools
  - Hotels, motels, and apartment buildings
- The preplan should include text as a descriptive purposes, a site plan, and a floor plan.
- The preplan should show or provide the following information:
  - Location
  - Construction features
  - Building emergency contacts
  - Occupancy type and load
  - Hazardous materials or processes
  - Location of utilities
  - Location of entrances and exits
  - Ventilation locations
  - Nearest water supply
  - Need for specialized extinguishing agents

- Private fire protection equipment and systems on site, including fire department connections
- Mutual aid plans
- In addition, some preplans provide for placement of apparatus on a first-alarm assignment

Although preincident planning components address many of the needs of responders to fires or hazardous material emergencies, preplanning also applies to EMS. For example, EMS responders may want to preplan alternate accesses to a nursing home after hours. Are there alternate means of egress? Will a stretcher fit in the building's elevator? What about access to locked apartments for a medical emergency when a person cannot make it to the door. Many times complexes have on-site maintenance or security with keys, which is easier and safer than breaking down the door and does not upset the owner as much.

Preincident planning is an excellent tool when considering preincident safety and risk identification. However, preincident planning is commonly limited to target hazards and many of the profession's injuries and deaths occur in single-family dwellings. In these cases preplanning can be useful in planned subdivisions where the dwellings are of the same construction and similar design.

SAFETY IN TRAINING

Based on the 2010 NFPA injury report, 7,275 firefighter injuries occurred during training. In terms of firefighter line-of-duty deaths, according to the 2010 USFA report, 12 firefighter deaths occurred during training activities. This was 13% of the total of 87 deaths. Training evolutions are generally created to simulate actual events, but the training evolutions must be better controlled so that injuries can be further prevented. *NFPA 1403, Standard on Live Fire Training Evolutions*, was created as a result of a number of injuries and deaths that occurred during live-fire training.

Following a tragic training accident in 1982 that resulted in the deaths of two firefighters, the NFPA developed the 1403 Standards in order to provide recognized safe practices for conducting such training evolutions. Unfortunately, even after the publication and adoption of *NFPA 1403*, training deaths during live-fire training continue to occur. To reduce these unnecessary deaths during live-fire training, the State of Florida has adopted rules for live-fire training that include the provisions of *NFPA 1403*. To conduct live-fire training evolutions in Florida, you need to be an Instructor I and hold the Live Fire Training Instructor I (LFTI-1) certificate. The 40-hour LFTI-1 class teaches you how to read and understand the *NFPA 1403* standard. During this class, students go out to an acquired structure to see if it meets *NFPA 1403* criteria. Most acquired structures need modifications, and students learn firsthand how to make the building compliant. Acquired structures may necessitate extensive modification, and you must decide if you are willing to do what it takes to make the structure compliant.

Training evolutions must be designed and conducted only after a thorough risk analysis that includes identification and control of the risks.

NOTE Injuries and deaths associated with training are avoidable, and the safety manager must adopt a zero tolerance level in this regard.

In *NFPA 1403,* Chapter 4 sets forth procedures and requirements for live-fire training in acquired structures. Chapter 4 addresses the following subjects:

- Student prerequisites
- Structures and facilities
- Fuel materials
- Safety
- Instructors

NFPA 1410, Standard on Training for Initial Emergency Scene Operations, describes the minimum requirements for evaluating the training of a fire department and the personnel engaged in emergency scene operations for initial fire suppression and rescue operations. It has been designed to provide fire departments with an objective method of measuring performance for initial fire suppression and rescue procedures using available personnel and equipment. Although not directly related to safety, using *NFPA 1410* can support safe practices by ensuring that evolutions can be completed safely using adequate numbers of personnel and proper equipment.

Many of the requirements addressed in these standards can become part of a department's standard procedures for training evolutions. Consider the use of protective clothing, for example—how often is it permitted that hose streams are operated with just helmets, or helmets and gloves, or worse yet, nothing? In this case, the training is not simulating actual conditions. Furthermore, the risk of injury from the hose is much greater than if the member was in full PPE. Is it department policy to assign a safety officer during a hazardous materials exercise, or during a multicompany drill using ladders at the drill tower? It should be policy. Do your policies provide for rehabilitation and EMS at the site of outside evolutions?

Do you follow the same safety procedures during EMS training as you would in the field in terms of disposal of blood products and needles? How about protective equipment during extrication exercises? Repeating cutting up cars in the summer in full PPE tends to make one cut corners. Having a safety officer watching and prohibiting such actions may save an injury.

NOTE The procedures must also provide for an accountability system at multicompany training evolutions similar to that used at real incidents.

This not only allows for practicing the system but also affords another level of safety for the participants.

EMPLOYEE WELLNESS AND FITNESS

An employee wellness and fitness program is an essential component of the safety and health plan. Fitness and wellness programs can be developed locally using in-house talent or by contracting with professional exercise and wellness consultants. A number of resources are available to assist departments in developing these programs.

The International Association of Fire Fighters (IAFF)/International Association of Fire Chiefs (IAFC)/American Council on Exercise (ACE) Peer Fitness Trainer (PFT) Certification Program is a part of the IAFF/IAFC Fire Service Joint Labor Management Wellness-Fitness Initiative. The purpose of the IAFF/IAFC/ACE PFT certification program is to provide a fitness trainer standard consistent with the health and fitness needs of the Fire Service throughout the United States and Canada.

The National Volunteer Fire Council launched the Heart-Healthy Firefighter Program in 2003 to combat the alarming trend of firefighter heart disease through education, awareness, and resources. The Heart-Healthy Firefighter Program promotes fitness, nutrition, and health awareness for all members of the fire and emergency services, both volunteer and career.

The USFA publishes a fitness coordinator's manual, local universities often have an industry hygienist who may be able to help, and department physicians may be able to provide information on program development. A number of components are necessary for a program to be successful and effective which are described throughout this text.

It is difficult to separate the terms *wellness* and *fitness*. Generally the occupational safety programs and health programs each have components that can fit in these categories, but for the programs to be effective, there must be tight integration. Wellness is a broad term that would include components relating to medical fitness, physical fitness, and emotional fitness. Wellness programs should also include behavior modification programs including weight reduction, tobacco use cessation, nutrition, and stress reduction.

A successful wellness and fitness program must also have provisions for after-injury rehabilitation and care. A rehabilitation program should be designed to ensure that the employee has a path to rehabilitation and return to duty with the employee's safety in mind. Rehabilitation programs should include input from the department doctor and physical therapy agencies familiar with the job of an emergency responder. **Transitional duty** programs are also helpful to permit a safe return. These programs allow an employee to work during his or her recovery, but within restrictions based on physician's orders. Periodic evaluations of the employee's condition are also an integral part of the program.

transitional duty Sometimes called alternative duty or modified duty, these programs allow an injured employee to return to work, with restrictions, for some period of time while recuperating; this allows the organization to reduce lost time while utilizing the skills of the injured member.

Medical Fitness

Medical fitness can be assessed by performing yearly medical exams. While recommendations differ as to how often these exams should be conducted for different age groups, it is agreed that annual medical exams are a necessity. *NFPA 1582* addresses the requirements for a comprehensive occupational medical program for fire departments. This standard requires a medical evaluation at the time of employment (prior to assignment) and annually thereafter. The medical exam should be conducted by a physician who is very knowledgeable in occupational issues and specifically with the essential job tasks, which are outlined in the standard. The standard provides valuable guidance for the fire department physician. The standard also identifies the essential tests and components of the medical exam. **Text Box 4-6** provides the general framework of the medical exams as recommended by the IAFF/IAFC wellness-fitness initiative, which are consistent with the requirements of *NFPA 1582*. *NFPA 1582* lists conditions for each body system and categorizes problems into either category A or B. Category A conditions are defined in the standard as those conditions that *would* preclude a person from performing tasks in a training or emergency operational environment because they would present a significant risk to the safety and health of the person or others. The fire department physician shall not certify for duty those persons with category A conditions. Category B conditions are defined in the standard as those that *could* preclude a person from performing tasks in a training or emergency operational environment by presenting a significant risk to the safety and health of the person or others, but do not automatically preclude a person from becoming certified as meeting medical standards for duty.

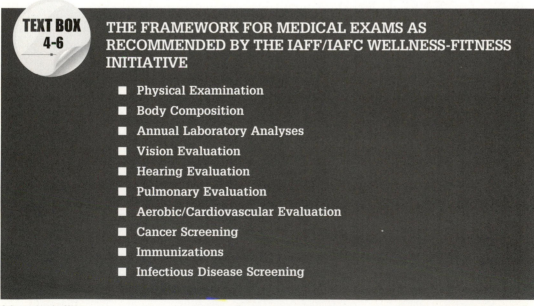

TEXT BOX 4-6

THE FRAMEWORK FOR MEDICAL EXAMS AS RECOMMENDED BY THE IAFF/IAFC WELLNESS-FITNESS INITIATIVE

- Physical Examination
- Body Composition
- Annual Laboratory Analyses
- Vision Evaluation
- Hearing Evaluation
- Pulmonary Evaluation
- Aerobic/Cardiovascular Evaluation
- Cancer Screening
- Immunizations
- Infectious Disease Screening

FIGURE 4-8
Ultrasound scanning
should be a part of
annual medical exams.

© Cengage Learning 2013

The preemployment and annual medical exams should also include ultrasound scans of the following body systems (See **Figure 4-8**):

- Echocardiogram Heart Scan
- Carotid Artery Scan
- Aortic Aneurysm Scan
- Liver, Pancreas, Gall Bladder, Kidneys, and Spleen (Abdominal Organ Scan)
- Ovaries/Uterus Scan
- Testicular/Prostate Scan
- Bladder Scan
- Thyroid Scan

Physical Fitness

According to *NFPA 1583, Standard on Health-Related Fitness Programs for Fire Fighters*, annual physical fitness assessments for emergency responders should address five components: **aerobic** or **cardiovascular fitness**, **muscular strength**,

aerobic fitness A measurement of the body's ability to perform and utilize oxygen.

cardiovascular fitness Fitness levels associated with the cardiovascular system, including the heart and circulatory system.

muscular strength The maximal amount of force a muscle or group of muscles can exert in a single contraction; the ability to apply force.

muscular endurance, **flexibility**, and **body composition**. These components interrelate and are all required of the emergency responder. A person can have a great deal of muscular fitness and be very strong, but have a low level of cardiovascular fitness or endurance. Clearly the work of an emergency responder requires both.

Cardiovascular fitness and body composition can be assessed as part of the medical exam. Muscular fitness can be assessed using a battery of tests. The handgrip dynamometer can be used to measure grip strength, the arm-flex dynamometer measures the maximum force generated from arm flexion, and the leg-extension dynamometer measures the maximum force that can be generated from extension of the legs. Push-ups and sit-ups can also be used to measure muscular strength and endurance.

Once the fitness level is identified, a program should be developed by the department fitness coordinator on an individual basis; certified peer fitness trainers can be of value in the development of these programs. For example, some employees may lack cardiovascular endurance, while others may have excellent cardiovascular endurance but lack muscular fitness. The program should also provide for on-duty time and the facilities for members to work out and maintain the required level of fitness.

Furthermore, the program must consider the following five basic components necessary for an exercise plan:

1. The mode or type of exercise
2. The intensity or difficulty of the exercise
3. The duration or length of the exercise session
4. The frequency or number of sessions per day or week
5. The progression, meaning the gradual increases in workload to promote a training adaptation

Following are the key considerations for fitness from the IAFF/IAFC wellness-fitness initiative:

- Medical clearance
- On-duty time for exercise
- Equipment and facilities
- Exercise specialists and peer fitness trainers
- Incorporating fitness throughout the Fire Service
- Fitness evaluations
- Exercise programs

muscular endurance The ability of the muscle to perform repeated contractions for a prolonged period of time; the ability of the muscle to persist.

flexibility The range of motion in a joint, which depends on the extensibility of soft tissues.

body composition A measure to show the percentage of fat in the body; there are certain published parameters for what is considered average or normal.

Emotional/Behavioral Fitness

Emotional or behavioral fitness is also a necessary component of the program. As part of the overall health and safety program, the following should be considerations for emotional fitness:

- Periodic behavioral health evaluation
- Comprehensive counseling services
- Incentive to access services
- Behavioral modification
- Employee/member assistance program
- Substance abuse
- Stress

A responder who is under undue stress, from whatever cause, will not be able to perform effectively and is prone to injury. Emotional fitness can be improved by having programs in place for employee assistance. These programs should be accessible to employees and their families on a confidential basis. Often employers pay the cost for a fixed number of visits through a contract with local behavioral care providers. Family members are encouraged to participate too, as often the emotional problems are related to family and home life. Services provided by some employee assistance providers include:

- Drug or alcohol abuse treatment
- Tobacco use cessation
- Family problems
- Financial problems
- Stress management
- Critical incident stress management

NOTE Employee assistance programs may also be part of the discipline process in an organization.

If an employee's work performance takes a sudden downturn, the reason may be an emotional problem. Requiring mandatory participation in an employee assistance program can be used as a step in the discipline process and can contribute to the future of the employee and the department.

Clearly a factor in the emotional wellness of an emergency responder is **critical incident stress management (CISM)**. Having a CISM program in place contributes to the health and safety program. This subject is presented in Chapter 8 with post incident safety issues.

critical incident stress management (CISM) A process for managing the short- and long-term effects of critical incident stress reactions.

INTERAGENCY CONSIDERATIONS

Another preincident safety and health issue that is sometimes overlooked is the relationship between the responder's agency and other agencies with which the organization cooperates. While these relationships have always had operational and safety implications, the tragic events in New York and Virginia on September 11, 2001, as well as the events that followed involving anthrax attacks, brought the need for interagency coordination to national attention. After these events, President Bush issued Homeland Security Presidential Directive 5, whose purpose was to enhance the ability of the United States to manage domestic incidents by establishing a single, comprehensive national incident management system (IMS).

In the forefront of interagency relationships is the relationship between mutual or automatic aid providers. In most emergency response agencies, some form of mutual/automatic aid agreements exists. The safety program manager must know how the other departments operate. Are they safety aware, do they use the same incident management system, the same accountability system? Do their operations comply with recognized standards? *NFPA 1710* requires certain operational issues to be included in mutual/automatic aid agreements. Are the communications systems compatible? Project 25 (P25), or APCO-25, refers to a suite of standards for digital radio communications for use by federal, state/province, and local public safety agencies in North America to enable them to communicate with other agencies and mutual aid response teams in emergencies. Once in place, this will allow responders to communicate with one another regardless of the radio system used. Having the ability to communicate information about hazards on an incident in a timely manner is essential for safe operations. Responders also frequently work with law enforcement agencies. In most cases, the emergency responders referred to in this text are not law enforcement officers. That creates the need for integration in response to some incidents and a preincident understanding of roles and relationships. Law enforcement agencies respond to many incidents with the fire or EMS responders including vehicle crashes, incidents requiring crowd or traffic control, incidents resulting from criminal activities, and in some jurisdictions as first responders to all medical incidents. In some areas, the law enforcement agencies may provide rescue or aeromedical evacuation. This response creates the need for integration in response to some incidents and a preincident understanding of roles and relationships (see **Figure 4-9**).

Having an understanding of each agency's priorities and roles prior to an incident occurring is critical to safety and health activities and must be preplanned. In some situations there will not be clear lines of authority and responsibility. In these situations, the incident management system must provide for **unified command**, which provides

unified command Used in the IMS when two or more jurisdictions or agencies share incident command responsibilities but do so in conjunction with each other.

FIGURE 4-9
At this vehicle crash, three different public safety agencies are working together.

Courtesy of Palm Harbor Fire Rescue

for an incident commander from both law enforcement and the emergency response agencies. These commanders work in unity to bring the incident to a safe conclusion. Other issues can be preplanned; for example, do the emergency responders wait for police clearance before entering a scene where violence such as shootings or stabbings is possible? Do law enforcement agencies provide for traffic control at an interstate accident, or is their priority to reopen the highway? What are the expectations of the emergency responder agencies in terms of protection provided by law enforcement during periods of civil unrest?

This relationship is a two-way street. Law enforcement may ask the emergency responders to avoid using reds lights and sirens when nearing a scene of a shooting in the street in order not to draw a crowd, making the officers' job unsafe. Or the aeromedical helicopter operated by the police may need some specific ground support for landing zones. All of these considerations can be dealt with on a local level and in many different ways.

NOTE The important point to realize is that the issues must be decided on prior to the incident and with the safety of *all* responders in the forefront of concern.

In terms of interagency cooperation, emergency responders will probably interact most with mutual aid and law enforcement, but other agencies are involved as well. Hospitals provide initial treatment or medical surveillance after exposures to infectious diseases. The relationship with the local emergency room should be an integral part of

FIGURE 4-10

Interagency incident management considerations—the command triangle—fire, law enforcement, and health/EMS—for expanded unified command.

safety and health programs. Again, what are the expectations of the emergency responder agencies and the expectations of the hospital? Depending on the department situation, a communication center may be operated by a separate agency wherein safety and health concerns for responders can be preplanned. The local health department can be a resource for information on infectious diseases. The health department's role has become more important to emergency responders as the threat of chemical, biological, radiological, nuclear, or explosive attacks on the country have increased. The health department has many resources available for these events, and in fact may be the lead agency depending on the incident. During a CBRNE incident, law enforcement, health, and the fire department may use a unified command as shown in **Figure 4-10**.

Depending on local conditions, safety and health for the emergency responder can be impacted, either positively or negatively, by a preincident relationship with agencies other than the responder's. This relationship can go far not only from a safety and health standpoint, but also from the standpoint of effective and integrated operations.

SUMMARY

Preincident safety encompasses a number of components including station safety, apparatus safety, response safety, preincident planning, safety during training, wellness/fitness programs, and interagency relations. Station and apparatus safety can be subdivided into issues relating to design and issues relating to ongoing, day-to-day operations. Response safety includes driver selection, driver training programs, and policies affecting response. Preincident planning is a tool that can be used by responders to effect safer operations. Building characteristics, occupancy, and particularly hazards can

be identified prior to an incident. Means of access and tactical considerations can also be discussed prior to an emergency and under better conditions.

Safety during training is a concern in the overall health and safety program. Training is often a simulation of real conditions and is a controlled, although somewhat unpredictable, environment. All training evolutions should be conducted with participant safety as the number one priority.

An integral part of the safety and health program is a program to ensure the highest level of fitness and wellness for the responders. Wellness is a broad term that encompasses a number of components, including medical, physical, and emotional fitness. Medical fitness can be assessed with medical exams. Physical fitness assessments are necessary and individual programs are developed for responder improvement. An employee assistance program is vital in ensuring emotional fitness.

Emergency responders work alongside and rely on a number of other agencies. Strong interagency coordination and preincident defined roles and expectation can be helpful in preventing injuries and death. Interagency coordination also leads to a more effective operation.

Concluding Thought: Safety is a state of mind. Many components can be prepared for before an incident.

REVIEW QUESTIONS

1. Which of the following would not normally be considered as part of the daily vehicle apparatus checks?
    a. Back flushing the pump
    b. Checking the emergency lights
    c. Checking the oil level
    d. Checking the seat belts
2. List the areas of concern when selecting emergency vehicle driver/operators.
3. Preventive maintenance programs should require biannual trips to the maintenance professional.
    a. True
    b. False
4. What is one option that exists for traffic light control at intersections?
5. List the components that make up the general framework that should be included in an annual medical examination.
6. Emotional fitness is one of three types of fitness important to the emergency responder.
    a. True
    b. False
7. List two external agencies with which a department may want to coordinate before an incident.
8. For the purpose of preincident planning, which of the following would not be considered a target hazard?
    a. A hospital
    b. A lumber yard
    c. A railyard
    d. A two-unit duplex

9. List three components of an effective vehicle preventive maintenance program.
10. List three methods of determining muscular fitness.

ACTIVITIES

1. Review your department's policy on medical examinations. Compare the policy to the standard and the recommendations in this text. How does your exam compare? If you do not have a medical examination program, research programs and write a draft policy.

2. Review your department's standard operating procedures for response to incidents. Is the policy current? Could the emergency response mode be reduced for some incidents? Is there a policy or procedure for the selection and training of drivers?

3. Review your department's standard procedures for safety in training. Does your live-fire training comply with *NFPA 1403*? With application of procedures and requirements from *NFPA 1403*, could improvements be made to day-to-day training activities?

4. Identify your department's relationship with other agencies. Are there standard procedures or other prearranged agreements? Are the policies in effect or do they need to be established?

CHAPTER 5

Safety at the Fire Emergency

LEARNING OBJECTIVES

Upon completion of this chapter, you should be able to:

- List the three incident priorities.
- Explain the relationship between the three incident priorities and their relationship to health and safety.
- Discuss in general terms the hazards faced by responders to fire incidents.
- List the components of personal protective equipment used for fire incidents.
- Discuss the need for and the components of an effective accountability system.
- Discuss the types of and the relationship between incident management systems and the health and safety of the responder.
- Describe the concept of and need for fireground air management.
- Define the need for and uses of a rapid intervention team.
- Discuss the components of incident rehabilitation.

Shortly after midnight on Sunday, April 12, 2009, a fire fighter and a fire captain were killed when they were trapped by rapid fire progression in a wind-driven residential structure fire. The victims were members of the first arriving company and initiated fast attack offensive interior operations through the front entrance. Less than six minutes after arriving on-scene, the victims became disoriented as high winds pushed the rapidly growing fire through the den and living room areas where interior crews were operating. Seven other fire fighters were driven from the structure but the two victims were unable to escape. Rescue operations were immediately initiated but had to be suspended as conditions deteriorated. The victims were located and removed from the structure approximately 40 minutes after they arrived on location.

NIOSH investigators concluded that, to minimize the risk of similar occurrences, fire departments should:

- Ensure that an adequate initial size-up and risk assessment of the incident scene is conducted before beginning interior fire fighting operations.

- Ensure that fire fighters and officers have a sound understanding of fire behavior and the ability to recognize indicators of fire development and the potential for extreme fire behavior (such as smoke color, velocity, density, visible fire, and heat).

- Ensure that fire fighters are trained to recognize the potential impact of windy conditions on fire behavior and implement appropriate tactics to mitigate the potential hazards of wind-driven fire.

- Ensure that fire fighters understand the influence of ventilation on fire behavior and effectively apply ventilation and fire control tactics in a coordinated manner.

- Ensure that fire fighters and officers understand the capabilities and limitations of thermal imaging cameras (TIC) and that a TIC is used as part of the size-up process.

- Ensure that fire fighters are trained to check for fire in overhead voids upon entry and as charged hoselines are advanced.

- Develop, implement, and enforce a detailed Mayday Doctrine to ensure that fire fighters can effectively declare a Mayday.

- Ensure fire fighters are trained in fireground survival procedures.

- Ensure all fire fighters on the fire ground are equipped with radios capable of communicating with the Incident Commander and Dispatch.

The type of duty that results in the largest number of firefighter fatalities is fireground operations. This emphasizes the need for strong fire operations–related policies and procedures as part of an organization's safety program.

Source: Information for this case study was taken in part from CDC/NIOSH Fire Fighter Fatality Investigation Report F2009-11. http://www.cdc.gov/niosh/fire/reports/face200911.html

INTRODUCTION

As seen in Chapter 1, most firefighter deaths and injuries occur while operating at fire scenes as a result of different causes, ranging from stress to building collapse. The safety and health program manager must have a good knowledge of the hazards faced at fires and procedures that will ensure safer operations.

Forming an understanding of the incident priorities and their relationship with operations and hazards will improve the safety manager's ability to design a training program, purchase equipment, and implement procedures targeted toward injury reduction.

Four of the 16 Firefighter Life Safety Initiatives have a direct or indirect relationship to safety at the fire emergency. These include:

1. *Define and advocate the need for a cultural change within the fire service relating to safety; incorporating leadership, management, supervision, accountability, and personal responsibility.*
2. *Enhance the personal and organizational accountability for health and safety throughout the fire service.*
3. *Focus greater attention on the integration of risk management with incident management at all levels, including strategic, tactical, and planning responsibilities.*
4. *All firefighters must be empowered to stop unsafe practices.*

INCIDENT PRIORITIES AND SAFETY

Priorities for incident management can be categorized and placed in descending order of seriousness into life safety, incident stabilization, and property conservation. These priorities are the first thing that the incident commander must consider in the incident, and they should be continuously reevaluated as the incident progresses as part of the risk management process.

Life Safety

Life safety is the first incident priority as it is based on the premise that life safety must be maximized.

NOTE Life safety is the group of activities that ensures that the threat of injury or death to civilians and emergency personnel is reduced to the absolute minimum.

This is done by limiting the exposure to danger as much as possible. In a simple situation, it may mean that emergency responders wear protective clothing while operating. In a complex incident involving a fire in several buildings, it may require the evacuation of several blocks of residences and setting up a collapse zone. Regardless of the situation or how the priority is met, life safety is always number one throughout the incident. Life safety activities must include and consider responder safety.

Incident Stabilization

The second priority is incident stabilization. During this time, attempts are made to solve the problem.

NOTE Incident stabilization is the group of activities required to stop additional damage or danger.

Stabilization could mean a quick interior attack on a room and contents fire or a defense attack on a fire to stop the spread from building to building. Size-up helps to determine initial strategy (see **Figure 5-1** and **Figure 5-2**).

FIGURE 5-1

Safety officer should perform a size-up based on incident priorities and communicate concerns to the incident commander and crews.

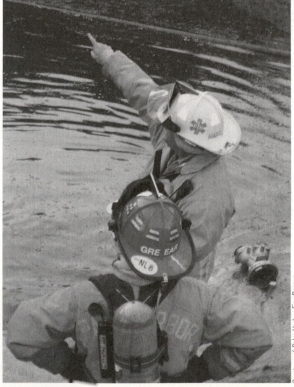

Courtesy of Palm Harbor Fire Rescue

FIGURE 5-2
Fire officer must perform a safety size-up on arrival at the incident scene.

Courtesy of Palm Harbor Fire Rescue

Property Conservation

Property conservation is the final incident priority and involves reducing the loss to property and the long-term health and welfare of the people affected. This group of activities is commonly called salvage or stopping the loss (see **Figure 5-3**). In a simple incident, the use of a salvage cover may be all that is necessary. In a large incident, the emergency response agency may be in property conservation mode for hours or days.

Relationship between Incident Priorities and Safety

Remember that life safety is always the number one priority and all operations must be developed based on this priority. As an incident priority, life safety applies to both civilians and emergency responders.

Recall from Chapter 3 that risk management began with risk identification. This first step, identification, should be the basis for risk management on the emergency scene. Because the life safety of the responders is the number one priority, incident operations and strategies can be developed that provide the highest level of safety to responders. In other words, for incident stabilization to occur, the incident commander should have considered life safety to the responders in formulating the strategy.

Risk assessment of the emergency scene can be based on the following philosophy:

We will take greater risk a lot to save a lot,

We will take little risk to save a little,

We will take no risk to save what is already lost.

FIGURE 5-3

Standard operating procedures must consider all phases of fireground activity. Incident commanders and incident safety officers must make a conscientious and informed decision as to what protective clothing is needed during overhaul.

Courtesy of Palm Harbor Fire Rescue

Although a risk analysis must always be performed, an incident commander would have few qualms about sending a search and rescue crew into a building fire if there were a possibility that people were inside. However, arriving at the same fire, meeting the occupants outside, and confirming that they are all out may produce a different strategy. Although some argue that a building is never unoccupied until a search proves it, more prudent and safety-minded incident commanders are changing strategies to make them more compatible with the risk.

The International Association of Fire Chiefs has published the rules of engagement for firefighter survival and for incident commanders. These are described in **Text Box 5-1.**

Hazards Faced by Responders

In order to do an effective risk assessment and to implement an effective strategy, it is necessary to understand the risks present at fires. For the purpose of this chapter, fire incidents are divided into four groups:

1. Structure fires
2. Transportation fires

TEXT BOX 5-1

RULES OF ENGAGEMENT FOR FIREFIGHTER SURVIVAL

1. Size up your tactical area of operation.
2. Determine the occupant survival profile.
3. DO NOT risk your life for lives or property that cannot be saved.
4. Extend LIMITED risk to protect SAVABLE property.
5. Extend VIGILANT and MEASURED risk to protect and rescue SAVABLE lives.
6. Go in together, stay together, come out together.
7. Maintain continuous awareness of your air supply, situation, location and fire conditions.
8. Constantly monitor fireground communications for critical radio reports.
9. You are required to report unsafe practices or conditions that can harm you. Stop, evaluate, and decide.
10. You are required to abandon your position and retreat before deteriorating conditions can harm you.
11. Declare a Mayday as soon as you THINK you are in danger.

The Incident Commander's Rules of Engagement for Firefighter Safety

1. Rapidly conduct, or obtain, a 360-degree situational size-up of the incident.
2. Determine the occupant survival profile.
3. Conduct an initial risk assessment and implement a SAFE ACTION PLAN.
4. If you do not have the resources to safely support and protect firefighters, seriously consider a defensive strategy.
5. DO NOT risk firefighter lives for lives or property that cannot be saved. Seriously consider a defensive strategy.
6. Extend LIMITED risk to protect SAVABLE property.
7. Extend VIGILANT and MEASURED risk to protect and rescue SAVABLE lives.
8. Act upon reported unsafe practices and conditions that can harm firefighters. Stop, evaluate, and decide.
9. Maintain frequent two-way communications, and keep interior crews informed of changing conditions.
10. Obtain frequent progress reports and revise the action plan.
11. Ensure accurate accountability of every firefighter's location and status.

(continues)

> 12. If after completing the primary search, little or no progress toward fire control has been achieved, seriously consider a defensive strategy.
>
> 13. Always have a rapid intervention team in place at all working fires.
>
> 14. Always have firefighter rehab services in place at all working fires.

Reprinted with permission of IAFC Safety Health and Survival Section. Copyright pending.

3. Outside structure fires

4. Wildfires

Fires related to specialized incidents such as hazardous materials are discussed in Chapter 7.

Structure Fires

Structure fires account for the majority of fires that injure and kill most firefighters. Depending on the makeup of a local community, these may or may not be the largest number of fires responded to. For discussion purposes, structure fires can be further subdivided, in terms of occupancy, into one- and two-family dwellings, multifamily dwellings, commercial, and industrial. **Table 5-1** compares hazards related to each of these occupancies. A similar

TABLE 5-1 Hazards Associated with Various Types of Occupancies

HAZARD	RESIDENTIAL 1 AND 2 FAMILY	MULTISTORY RESIDENTIAL	COMMERCIAL	INDUSTRIAL
Life hazard	High	High	Medium, depends on time of day	Medium, depends on time of day
Structure failure	X	X	X	X
Large area operations		X	X	X
Exposures threatened by fire		X	X	X
Back draft	X	X	X	X
Flashover	X	X	X	X
Explosions			X	X
Chemical	X	X	X	X
Improper storage	X	X	X	X
Long operations/need for rehabilitation		X	X	X

© Cengage Learning 2013

table could be completed after a risk identification process in your jurisdiction. The hazards listed are general enough to apply to buildings of different sizes, configurations, and heights. However, a number of hazards are common to any structure fire. Some hazards are directly related to the structure in terms of construction.

NOTE Risk assessment should consider the construction type and the anticipated reaction of the construction methods and materials to fire and heat.

For example, it is no secret that because of their design, lightweight, wood-truss roof assemblies have failed early in operations, and the incident commander should consider this in the risk analysis. Hazards relating to the structure that should be considered are:

- Types of loads and transmission of loads (fixed, live, dead, etc.)
- Construction type
- Building material types
- Structural elements including beams, girders, columns, connectors, and support systems
- Avenues for fire and smoke spread

   Again, risk identification is necessary to identify local risks.

   Because many structure fires are extinguished from the inside of the building, fire behavior characteristics must also be considered. The incident commander and safety manager must be familiar with the signs and potential outcome of situations such as **back drafts**, **flashovers**, **mushrooming**, **rollover**, and **heat transfer** (see **Figure 5-4**). Through training, education, experience, and the proper strategic and tactical coordination of fire attack and ventilation, the incident commander can make risk assessments and minimize the exposure of responders to these potentially deadly hazards.

back drafts Occur as a result of burning in an oxygen-starved atmosphere; when air is introduced, the superheated gases ignite with enough force to be considered an explosion.

flashover A sudden, full involvement in flame of materials and gases within a room.

mushrooming Heat and gases accumulate at the ceiling or top floor of a multistory building and then back down; can be prevented with vertical ventilation.

rollover The rolling of flame under the ceiling as a fire progresses to the flashover stage.

heat transfer The transfer of heat through conduction, convection, radiation, and direct flame contact.

Transportation Fires

Fire involving transportation includes vehicle fires (both auto and truck), trains, aircraft, and ships. Each of these forms of transportation presents unique problems to the

FIGURE 5-4
Proper tactics, including ventilation, reduce the chance of back drafts.

Courtesy of Palm Harbor Fire Rescue

emergency responders (see **Figure 5-5**). Each form of transportation fire could make a textbook by itself. However, like structure fires, some generalizations can be made and risk identification can assist in assessment of the local risk.

Generally, hazards associated with vehicle fires involve the vehicle itself, the fuel used to power the vehicle, and what cargo the vehicle is carrying. Fire involving the vehicle components themselves may weaken structural components and allow them to fail. When they fail under heat or fire, some vehicle components may explode or become flying missiles. For example, some bumpers on vehicles that are designed to withstand low-speed collisions without damage work by using a piston. When exposed to heat, the gases in the piston expand and ultimately the piston can fail, sending the bumper hurling toward responders. The fuel that powers a vehicle is a flammable or combustible liquid, which adds to the fuel of the fire. Modern hybrid vehicles that are powered by both fuel and electricity present unique hazards as well. Gasoline is stored in a traditional fuel tank while electricity is stored in a high-voltage battery pack. Hybrid vehicle manufacturers have published detailed safety manual for responders. The NFPA Electric Vehicle Safety Training project is providing firefighters and first responders with the information and materials necessary to respond to emergency situations involving electric vehicles. This training will help first responders identify electric vehicles and respond to common hazards.

Information from these manuals and training can assist the health and safety officer in developing training and standard operating procedures for these types of vehicle hazards.

FIGURE 5-5
Transportation fires can range from the family car to large trucks, railcars, airplanes, ships, or small water craft.

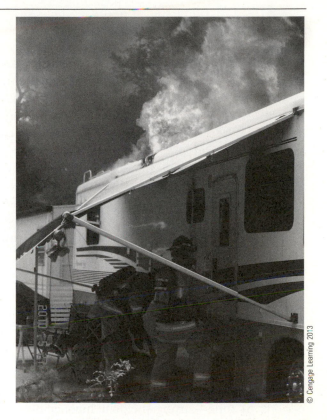

© Cengage Learning 2013

Also, although usually thought of only in the case of trucks, the cargo itself is of concern in any vehicle accident. How many people take their small barbeque tank to the refill station in the trunk of their car? Or how many travel home from the pool store with pool chemicals?

NOTE Vehicle fires large or small require the same risk assessment and application of incident priorities as do structure fires.

Transportation fires most often occur along a roadway, so traffic management is also a concern. Because most highway incidents involve vehicle crashes, this subject is discussed in detail in Chapter 6. On-scene risk assessment is necessary to ensure the application of appropriate strategies.

Outside Structure Fires

Fires occurring outside of structures can range from dumpster fires to grass fires. Each can present different levels of hazards and risks.

NOTE Dumpster or garbage fires require special attention, as the responder normally has little information regarding the contents of the containers.

Some clues can present themselves, however. A dumpster behind a hardware store can lead one to assume that paints and other chemicals might be involved. Other possible contacts in dumpster fires include hazardous materials and bloodborne pathogens. One fire fighter was killed and eight fire fighters were injured when the contents of a dumpster exploded at a foundry in Wisconsin. Proper protective equipment, including SCBA, and strong risk management are essential at dumpster fires. Further information can be found in the NIOSH Fire Fighter Fatality Investigation and Prevention Program Report # F2009-31.

Outside fires also expose responders to certain topographic and environmental hazards, including weather and access hazards. These fires often do not have a civilian life hazard associated with them, so a good risk assessment in these situations is "risk a little to save a little." This philosophy can go a long way in preventing injuries and fatalities.

Wildfires

Some of the most publicized fires are wildfires. Whether the fire is in a forest in Wyoming or in the urban interface of California, these fires can be both large and long in duration, and can produce enormous property and environmental loss. They also are responsible for firefighter injuries and deaths. Some metropolitan-area fire departments never see a wildfire, while for some departments, wildland fires are very common. Because of the nature of these fires—commonly over large geographic areas and for long periods of time—strong incident management systems with good accountability and communication procedures must be in place (see **Figure 5-6**).

FIGURE 5-6
Wildfire in progress. Note the onlookers—another safety concern for responders.

© Cengage Learning 2013

TEXT BOX 5-2

SAFETY TIPS FOR WILDLAND FIRES

Ten Firefighting Orders

1. Keep informed on fire weather conditions and forecasts.
2. Know what the fire is doing at all times.
3. Base all actions on the current and expected behavior of the fire.
4. Plan escape routes for everyone, and make them known.
5. Post a lookout where there is possible danger.
6. Be alert, keep calm, think clearly, act decisively.
7. Maintain communication with your crew, your supervisor, and adjoining forces.
8. Give clear instructions, and be sure they are understood.
9. Maintain control of your crew at all times.
10. Fight fire aggressively, but provide for safety first.

© Cengage Learning 2013

The crews fighting these fires are often very far from help, so it is critically important that the incident management system provides for long-term planning and logistical needs, such as food and rehabilitation of crews. Safety program managers and incident commanders who function in wildfire-prevalent areas must have knowledge of the weather and its effect on the fire. Included in the concept of weather are the effects of temperature, air masses, temperature inversions, relative humidity, wind velocity direction, and fuel moisture. For further risk assessment, consideration must be given to topography, including elevation and slope; to fuel, including ground fuels, aerial fuels, and structural fuels; and to fire behavior. These incidents take the incident management system to high levels of organization, often filling positions that were only designed for special wildland situations. **Text Box 5-2** discusses safety tips for wildland firefighting.

PERSONAL PROTECTIVE EQUIPMENT

Personal protective equipment (PPE) designed for firefighting is one strategy for minimizing exposure at the fire scene and addressing the priority of life safety as applied to responders. PPE for fighting structure and vehicle fires is essentially the same. Wildland firefighting has a different PPE design; although many departments do provide specialized PPE for wildfires, many do not. Specialized PPE for incidents such as technical rescue and hazards materials is discussed in Chapter 7.

A discussion on PPE for response to fires can be divided into three subject areas:

1. Design and purchasing
2. Use
3. Care and maintenance

Standards from the NFPA set forth requirements for both the purchase and the care and maintenance of firefighting PPE. Specifically, *NFPA 1851* specifies the minimum selection, care, and maintenance requirements for structural firefighting protective ensembles; and the individual ensemble elements including coats, trousers, coveralls, helmets, gloves, footwear, and interface components must be compliant with *NFPA 1971*. *NFPA 1971* specifies criteria and test methods for protective clothing designed to protect firefighters against environmental effects during structural firefighting. *NFPA 1852* specifies the minimum requirements for the selection, care, and maintenance of open-circuit self-contained breathing apparatus (SCBA) and combination SCBA/supplied-air respirators (SAR) that are used for respiratory protection during emergency operations in environments where the atmosphere either is Immediately Dangerous to Life and Health (IDLH) or could become oxygen deficient or IDLH.

NFPA 1500 requires that emergency responders be provided with PPE based on the hazards of the particular work environment. Once provided, guidelines or procedures must be adopted governing both the use of the PPE and a care and maintenance program, including inspections.

Design and Purchase

The *NFPA 1500* standard requires that new firefighting PPE meet the current editions of the respective standards. Older gear must have met the standard in effect when purchased. Therefore, the purchasing agent for the department should reference the particular NFPA standard, in its entirety, in the specification for the PPE. From a safety and health standpoint, the specification writing process should include input from the safety committee or, at a minimum, from the safety program manager.

The commonly expected components of PPE for firefighting (see **Figure 5-7**) include:

- Approved fire helmet with protective eye shield
- Flame-resistant hood
- Turnout coat
- Turnout pants
- Firefighting gloves
- Firefighting boots
- PASS devices
- SCBA[1]

FIGURE 5-7
Firefighters in full NFPA-compliant firefighting
PPE.

© Cengage Learning 2013

The safety program manager must be involved in the purchasing of firefighting
PPE to ensure that the PPE meets the applicable standards. To be effective, the safety
program manager must also be familiar with the requirements of the standards and
their implications for firefighter safety. For example, the safety program manager
must understand the relationship between **thermal protective performance
rating (TPP)** and **total heat loss (THL)**. When considered together, these values
produce a good indication of the level of safety and performance provided by the PPE
ensemble.

thermal protective performance rating (TPP) The amount of protection against both convective and radiant heat that your composite gear outer shell, thermal liner, and moisture barrier should be able to provide in the event of a flashover.

total heat loss (THL) The measurement of an ensemble's heat stress reduction capability; the higher the THL value, the greater the benefit for the firefighter.

TEXT BOX 5-3

EXAMPLE PPE USE PROCEDURE

Type of Incident	Protective Clothing Required
Smoke checks in area	Full bunker gear
Wires down, electrical arcing, overheated appliances, etc.	Full bunker gear
Vehicle accidents	Full bunker gear for those involved in hazardous activities. Patient care personnel, at the discretion of the officer in charge
Performing extrication/heavy rescue	Full bunker gear
Grass or brush fires	Full bunker gear or Wildfire PPE (except brush truck driver, who shall have his/her gear readily available). If needed, SCBA
Structure fires & automatic fire alarms	Full bunker gear and SCBA
Smoke checks in structure	Full bunker gear and SCBA
Hazardous liquid spills (suspected fuel spill)	Full bunker gear and SCBA until product is identified. Full bunker gear and appropriate respiratory protection thereafter, as determined appropriate
Vehicle & dumpster or trash fires	Full bunker gear and SCBA
Gas leaks or investigate possible hazardous materials	Full bunker gear and SCBA
Overhaul and salvage	Boots, bunker pants, gloves, and helmets. SCBA until thoroughly ventilated

Source: Pinellas County Fire Chiefs' Association, Operations Chiefs Committee

Use

SOPs defining the use of the PPE should also be in place. Although this seems quite obvious, it is helpful from the safety standpoint to require the use of full PPE under specific conditions. Examples might include that full PPE must be used for interior structure firefighting. But when is it permissible to remove the SCBA? During overhaul? Studies have shown that continued SCBA usage during overhaul is essential. See **Text Box 5-3** for an example policy of PPE use as it relates to fire emergencies.

The SCBA policy should require that the SCBA mask be on until the safety officer or incident commander deems the atmosphere in the work area to be safe. This decision can only be safely made by using air-monitoring equipment (see **Text Box 5-4**).

TEXT BOX 5-4

EXAMPLE SCBA USE PROCEDURE

SCBA Use and Safety

1. SCBA shall be used by personnel working in areas where:

   A. Personnel are working below ground or inside any confined space unless the atmosphere can be established by testing and continuous monitoring.

   B. Airborne combustible dust at or above its LEL is present, or less than 5 ft. of visibility.

   C. Atmospheric concentration of any substance at or exceeding its Time Weighted Average (TWA),* or any other atmospheric condition that is Immediately Dangerous to Life or Health (ILDH).
   The chart below lists the "most commonly" monitored fire ground gases and does not represent all atmospheric dangers.

TIME WEIGHTED AVERAGE (TWA) VALUES	
Oxygen (O_2)	19.5% (low) 23.5% (high)
Hydrogen Cyanide (HCN)	4.5 ppm
Hydrogen Sulfide (H_2S)	10 ppm
Carbon Monoxide (CO)	35 ppm
Lower Explosive Limit (LEL)	10%

*TWA is the concentration of a contaminant averaged over a workday (usually 8 hours long). It's measured in a workplace by sampling a worker's breathing zone for the whole workday.

2. Use of SCBA shall be in conjunction with department-issued protective gear mentioned previously.

3. Removal of SCBA within a previously hazardous atmosphere shall be at the direction of the officer in charge of interior operations or the Safety Officer. Removal of SCBA outside the hazardous area will be at the discretion of the company officer.

4. Personnel using SCBA shall perform the negative and positive pressure mask fit test prior to each entrance into the hazardous atmosphere.

5. PASS device shall be in place and activated with all uses of SCBA.

6. Personnel using SCBA shall operate in teams of two or more who are always in communication with each other through visual, audible, physical, safety guide rope, electronic or other means to

(continues)

coordinate their activities, and are in close proximity to each other to provide assistance in the event of an emergency.

7. It is recommended that initial and final atmospheric readings be communicated to dispatch for inclusion in incident notes. This is done via radio communication or use of Mobile Data Terminals.

Source: Pinellas County Fire Chiefs' Association, Operations Chiefs Committee

Even after the area is deemed safe from the air monitoring perspective, there must be consideration for continued respiratory protection due to dangerous particles of combustion. The health and safety officer must consider procedures for proper PPE during the fire investigation phase of the fire.

Procedures can also define different levels of PPE for different types of fires. Is SCBA required for wildland fires? How about a dumpster fire? A firefighter would never approach a hazardous materials incident without wearing SCBA, yet would fight a dumpster fire behind a hardware store without SCBA. The use of PPE can be linked to common sense. However, like other components of a complete safety program, the proper use of PPE should be outlined in policies and procedures.

Care and Maintenance of PPE

The NFPA standards on PPE also provide guidelines for the care and maintenance of the equipment.

NOTE The safety program must have a component that includes PPE inspection procedures, repair procedures, and care.

Generally, the standards require following the manufacturer's recommendations.

Because the PPE is issued to individual members, often the care is also delegated to them. However, because this equipment is specialized, the department has an obligation to provide a process or the facilities for cleaning and minor repairs per standard practices. In some departments, both of these functions are handled in-house, while other departments contract with outside vendors to provide this service. If the department also has EMS responsibilities, the bloodborne pathogens procedures must also be considered in the cleaning of the PPE.

Although the care is delegated to the user, the safety program must include a procedure for periodic inspection to ensure the PPE is in good condition. Monthly inspections are recommended (see **Figure 5-8**). The monthly inspection can be performed by a station officer or, in some departments, the shift commander. However, the safety program manager should also be involved in the process. A quarterly or semiannual

FIGURE 5-8
Company officer doing monthly inspection of PPE.

inspection by the safety program manager or staff is a good idea. After each use, the PPE should be inspected by the user. Of course, SCBA and PASS devices should be checked and inspected at the beginning of each shift and after use. **Figure 5-9** provides an example of a PPE inspection form.

Records should be kept on all PPE. The record should include assignment, dates when cleaned, date and type of repairs, and inspection dates. This record should be filed in the workplace of the person who has the PPE assigned to him or her. **Table 5-2** is an example of a PPE record.

INCIDENT MANAGEMENT SYSTEMS

A key requirement to maximize safety of the fire scene is the use of an **incident management system (IMS)**. Further, *NFPA 1561, Standard on Emergency Services Incident Management System*, provides the requirements of an incident management system. A number of recognized systems are in use throughout the country. Local conditions and procedures often dictate the type of system but, to be effective, the system must have the following design characteristics:

- Must provide for different kinds of operations, including single jurisdiction/single agency response, single jurisdiction/multiagency response, and multijurisdiction/multiagency response.
- Organizational structure must be adaptable to any emergency.
- Must be usable and understood by all agencies in a particular geographic area.

incident management system (IMS) An expandable management system for dealing with a myriad of incidents to provide the highest level of accountability and effectiveness; limits span of control and provides a framework of breaking the big job down into manageable tasks.

EXAMPLE
PERSONAL PROTECTIVE EQUIPMENT RECORD

Employee Name: _____   Year: _____

Item	Inv #	Issued	July		Aug.		Sept.		Oct.		Nov.		Dec.	
Helmet														
Hood														
SCBA Mask														
Mask Bag														
Safety Glasses														
Bunker Coat														
Gloves														
Bunker Pants														
Suspenders														
Bunker Boots														
Work Boots														
HEPA Mask														
////////			Inp/Res		Inp/Res		Inp/Res		Inp/Res		Inp/Res		Inp/Res	

Inp=Inspection Date and Inspector Initials/ Res=Result of Inspection (Pass or Fail)

REMARKS

July:_____
_____

August: _____
_____

September:_____
_____

October:_____
_____

November:_____
_____

December:_____
_____

© Cengage Learning 2013

FIGURE 5-9

A typical inspection check sheet.

TABLE 5-2 Sample PPE Record

ASSIGNED TO:
STATION #:
ID#:
SHIFT:

ACTIVITY	HELMET	COAT	PANTS	GLOVES	HOOD	BOOTS	PASS	SCBA	REMARKS (USE BACK AS NEEDED)
6 Month Clean 1/1/2012	X	X	X	X	X	X			
Inspection 1/5/2012	X	X	X	X	X	X	X	X	Pass battery due Hole left pant knee
Knee repaired 1/15/2012			X						
Battery replace 1/15/2012							X		
Annual Flow Test								X	Completed by Smith's SCBA shop

FIGURE 5-10
Incident management
system—command staff
and general staff positions.

- Must be expandable in a logical manner from the beginning of an incident.
- Must use common terminology.
- Must have integrated communications.
- Must maintain a manageable span of control.
- Must be used on all incidents!

Because this is not a text on IMS, the focus here is on the role of safety within the IMS structure. The IMS structure provides for a command staff and general staff. Included in the command staff are three functions or positions: information, safety, and liaison. The general staff has four positions: operations, finance, logistics, and planning (see **Figure 5-10**). Each of these functions is performed at each and every incident. At a single unit response incident, the company officer may fill all the command staff roles; at a one-alarm house fire the shift commander may fill all the roles except for safety and information. At a multialarm commercial fire, administrative officers may fill in many of the command staff or general staff functions. At a several weeks–long fire fight at a wildfire, all IMS positions may be filled and the system may be several layers deep.

From a safety and health perspective, the incident safety officer is part of the command staff.

NOTE To be effective, the safety officer must have expertise in fireground risk management.

The safety officer must also have authority within the IMS and direct communication with the incident commander. Requirements and roles of the incident safety officer are further described in Chapter 9.

CHAPTER 5
SAFETY AT THE FIRE EMERGENCY 129

NOTE At an incident scene, the safety officer or a member of the safety sector or group must be given certain authorities.

These include stopping, altering, and suspending operations that are determined to be unsafe. The safety officer must alert the incident commander of the situation and what action the safety officer has taken. The safety officer working at a fire scene should be alert to a number of safety-related actions, proper PPE being worn, structural conditions, establishing collapse danger zones, overseeing the accountability function, preventing **freelancing**, and ensuring that provisions are made for responders' rehabilitation.

freelancing Occurs when responders do not follow the incident plan at a scene and do what they want on their own; a failure to stay with assigned group.

AIR MANAGEMENT

In fire responses that require use of self-contained breathing apparatus and require the responder to enter the hazard area, air management policies should be established. Studies have indicated that a 30-minute SCBA bottle does not last 30 minutes. Waiting for the alarm bell to sound may not give enough time to safely exit the hazard area.

Therefore, the health and safety program must including training and procedures or rules of air management. The components of these rules include:

- Understanding each firefighter's average air consumption. This can be accomplished through training drills.
- Emphasizing that the firefighter must know the amount of air they have left and must leave the hazard area before the low pressure alarm sounds.
- Emphasizing the need to have enough air remaining for any unplanned events, such as disentanglement or disorientation, as you exit the hazard area.

ACCOUNTABILITY

Personnel accountability, knowing who they are, where they are, and what they are doing, is a critical element of the IMS and of the safety program. Both *NFPA 1500* and *NFPA 1561* require accountability systems at the emergency scene and both place the responsibility for accountability with the incident commander. Accountability occurs by using several layers of supervision in various geographic areas or functional areas. Each supervisor is responsible for his or her operating crews with the concept of company or **crew unity** being of paramount importance. An accountability officer should

crew unity The concept that a fire company or unit shall remain together in a cohesive, identifiable working group to ensure personnel accountability and the safety of all members; a company officer or unit leader shall be responsible for the adequate supervision, control, communication, and safety of members of the company or unit.

FIGURE 5-11
The typical passport and accountability board.

be assigned within the IMS system or, on smaller incidents, assignments could be handled by the incident commander or safety officer.

As with the IMS, there are several variations in accountability systems in use. Some use the passport system, in which the crew's names are placed on small Velcro tags and then placed on a small card that is used at the incident scene as a passport (see **Figure 5-11**). Others use two-dimensional bar codes and computers. Yet another system uses a card and clip that are attached to a large ring carried in the apparatus. Systems can be purchased commercially or designed locally, but the accountability system should meet the following objectives:

- Account for the exact location of all individuals at an emergency scene at any given moment in time.
- Provide for expanding to meet the needs of the incident.
- Be adaptable to the IMS in use.
- Ensure that all individuals are checked into the system at the onset of the incident.
- Provide for visual recognition of assignment.
- Provide for points of entry into the hazard zone.
- May also include medical data and the individual's training data.

NOTE Locally adopted accountability procedures are a strict requirement and oblige participation by everyone at all levels and at all incidents.

For an accountability system to be effective, it must have support by senior management of the organization. Each level has certain responsibilities to the system. Individual responders are responsible to ensure that their name tags are in the proper location on the apparatus and at the scene it is their responsibility to stay in direct contact with their assigned crew. *No freelancing!* Officers are responsible for knowing the location of each person assigned to them and for ensuring that they stay within their assigned work area. Division or group supervisors are responsible for the accountability within the work area. Accountability officers are responsible for the accountability of any area of the incident from a specific point of entry. Incident command is responsible for putting the accountability function into the IMS and to provide a means for accounting for every individual at the scene. An accountability division/sector/group at a large incident may determine the points of entry, ensure communication with accountability officers, and provide the overall coordination of accountability at a specific incident.

The accountability system requires **personnel accountability reports (PARs)**. These PARs require the incident commander or accountability officer to check the status of crews and ensure that all personnel are accounted for during certain regular intervals or benchmarks during the incident. These benchmarks should be defined and may be related to time of activity. Examples of benchmarks include:

- Fixed time, generally every 30 minutes
- After primary search
- After fire is under control
- After a switch in strategy (offensive to defensive)
- After a significant event such as a collapse, flashover, back draft
- After any report of a missing firefighter

personnel accountability reports (PARs) A verbal or visual report to incident command or to the accountability officer regarding the status of operating crews; should occur at specific time intervals or after certain tasks have been completed.

RAPID INTERVENTION COMPANIES

Regardless of the effectiveness of the IMS in place, or the greatest of accountability systems, nothing is gained if personnel are not available to respond when an unexpected event occurs at the emergency scene. The IMS and accountability system can provide the incident commander with the information that a crew is missing, but to intervene, the incident commander must have the resources close by to handle the situation. This is the concept of **Rapid Intervention Companies (RIC)** or Teams (RIT). Although

rapid intervention companies (RIC) An assignment of a group of rescuers with the sole purpose of rapid deployment to reports of operating personnel in trouble or missing.

called by many other names, the RIC is a group of firefighters—be it two, three, four, or more—who are fully equipped and have complete PPE on, and who stand by near the emergency scene and wait for something to happen. When something does happen—whether it is a lost or trapped firefighter or injured firefighters requiring assistance—this crew is ready to respond and assist. As described in Chapter 1, the research in Phoenix clearly revealed that it is necessary for the fire department to practice rapid intervention drills and skills. It was apparent after 200 drills that much more can be learned regarding the use of rapid intervention companies. It is the responsibility of the safety program manager and the training staff of the organization to ensure the rapid intervention company is prepared to be effective and is not just an entry on the incident command worksheet.

Although the RIC would meet the requirements of the two-in/two-out rule introduced in Chapter 2, there is ongoing confusion and debate within the fire service as to whether there is a difference—conceptually or functionally—between the two-in/two-out rule and rapid intervention companies. Some fire departments consider the two to be synonymous, that is, the terms can be used interchangeably. Other departments view RICs as a higher order two-in/two-out team that is differentiated by specialized training (see **Text Box 5-5**) and equipment, and possibly even by the firefighting credentials of RIC members.

For the purposes of this text, the two-in/two-out requirements pertain to initial arrival of units, which is referred to in *NFPA 1710* as the initial RIC. This initial RIC uses the initial arriving units and is deployed as part of first-due operations. Once a scene progresses beyond the incipient stage and escalates to a working fire or additional alarm or both, then the need to deploy a formal RIC or RICs becomes a factor. Firefighter safety is the issue with both RIC and two in/two out. Each has its own requirements and distinct, related approaches to providing safety based on the level of need. Both are implemented and often staffed differently, but are established with the common goal of effecting firefighter rescue.

Rapid intervention crews can be implemented in a number of ways. Some departments dispatch an extra unit to working fires to function as this team; others dispatch a company on the first alarm; for example, the fourth-due engine might function as the RIC. Some use rescue or advanced life support units, others have designated teams such as special operations that respond to all structure fires.

The number of RIC necessary at an emergency incident can vary with the complexity of the incident. A room and contents house fire would probably only require a crew of two or three, while a large warehouse complex may require a RIC on each side of the building, simply because of distance.

On arrival, the team should size up the building and the operation and try to anticipate what could happen. It must determine possible entry and exit points. The tools should be assembled for the RIC to use based on the type of building. For example, in a two-story garden apartment building, the RIC may want to have a 24-foot ladder with it

TEXT BOX 5-5

SUGGESTED TRAINING TOPICS FOR RIC PERSONNEL

- Building construction
- Rescue scenarios:
  - Entanglement
  - Floor collapse
  - Confined space
  - Ground level
  - Above ground
  - Below ground
- Incident size-up
- Fire behavior and travel
- Team search techniques and problems (including large-area searches)
- Use of thermal imagers
- SCBA changeover and use of emergency breathing support systems
- Self-rescue techniques
- Fireground communications—frequency and capability
- Forcible exit
- Accountability
- Ladder bail out
- Methods of firefighter removal
  - Unconscious
  - Conscious
- Rope/charged hose line slide
- Command of RIC operations

Source: USFA/FEMA

for rapid access to the second floor; in a large metal-sided warehouse, the RIC should have a power saw with a metal blade. **Text Box 5-6** lists the minimum equipment that should be assembled and carried by the RIC.

This RIC concept has widespread applicability, not only to fire emergencies but also to other incidents as well. The incident commander must provide for a RIC early in the incident and throughout the duration.

TEXT BOX 5-6

MINIMUM EQUIPMENT FOR A RIC

- Extra SCBA complete with harness, regulator, and extra masks (consider that automatic/mutual aid companies may use different SCBA systems)
- Search rope
- Forcible-entry hand tools such as axe, sledge, halligan bar, and bolt cutters
- Mechanical forcible entry tools such as chain saw, metal-cutting saw, and masonry-cutting saw
- Hose line available
- Ladder complement (size and type depends on the building height)
- Thermal imaging camera
- High-intensity hand light

Source: USFA/FEMA

REHABILITATION

The physical and mental demands placed on responders coupled with the environmental dangers of extreme heat and cold will have an adverse effect on the responder from a safety perspective. **Figure 5-12** and **Figure 5-13** show the heat stress index and the wind chill index, respectively. Crews that have not been provided with adequate rest and **rehabilitation** during a fire or other emergency operation are at increased risk for illness and injury.

NOTE Rehabilitation on the emergency scene is an essential element of the IMS (see **Figure 5-14**).

This need for rehabilitation is also cited in a number of the national standards relating to safety and fire scene operations. *NFPA 1584, Recommended Practice on Rehabilitation of Members Operating at Incident Scene Operations and Training Exercises,* applies to personnel operating at both emergency scenes and training exercises and requires that the IMS have a component for rehabilitation.

The development of an incident rehabilitation program has minimal impact on an organization and therefore should be achievable. While systems can be very different, the basic plan should provide for the following:

- Establishment of a rehabilitation sector/group within the IMS
- Hydration

rehabilitation The group of activities that ensures responders' health and safety at an incident scene; may include rest, medical surveillance, hydration, and nourishment.

RELATIVE HUMIDITY

TEMPERATURE °F	10%	20%	30%	40%	50%	60%	70%	80%	90%
104	98	104	110	120	132				
102	97	101	108	117	125				
100	95	99	105	110	120	132			
98	93	97	101	106	110	125			
96	91	95	98	104	108	120	128		
94	89	93	95	100	105	111	122		
92	87	90	92	96	100	106	115	122	
90	85	88	90	92	96	100	106	114	122
88	82	86	87	89	93	95	100	106	115
86	80	84	85	87	90	92	96	100	109
84	78	81	83	85	86	89	91	95	99
82	77	79	80	81	84	86	89	91	95
80	75	77	78	79	81	83	85	86	89
78	72	75	77	78	79	80	81	83	85
76	70	72	75	76	77	77	77	78	79
74	68	70	73	74	75	75	75	76	77

NOTE: Add 10°F when protective clothing is worn and add 10°F when in direct sunlight.

HUMITURE °F	DANGER CATEGORY	INJURY THREAT
BELOW 60°	NONE	LITTLE OR NO DANGER UNDER NORMAL CIRCUMSTANCES
80°–90°	CAUTION	FATIGUE POSSIBLE IF EXPOSURE IS PROLONGED AND THERE IS PHYSICAL ACTIVITY
90°–105°	EXTREME CAUTION	HEAT CRAMPS AND HEAT EXHAUSTION POSSIBLE IF EXPOSURE IS PROLONGED AND THERE IS PHYSICAL ACTIVITY
105°–130°	DANGER	HEAT CRAMPS OR EXHAUSTION LIKELY, HEAT STROKE POSSIBLE IF EXPOSURE IS PROLONGED AND THERE IS PHYSICAL ACTIVITY
ABOVE 130°	EXTREME DANGER	HEAT STROKE IMMINENT!

FIGURE 5-12
Heat stress index.

TEMPERATURE °F

WIND SPEED (MPH)	45	40	35	30	25	20	15	10	5	0	-5	-10	-15
5	43	37	32	27	22	16	11	6	0	-5	-10	-15	-21
10	34	28	22	16	10	3	-3	-9	-15	-22	-27	-34	-40
15	29	23	16	9	2	-5	-11	-18	-25	-31	-38	-45	-51
20	26	19	12	4	-3	-10	-17	-24	-31	-39	-46	-53	-60
25	23	16	8	1	-7	-15	-22	-29	-36	-44	-51	-59	-66
30	21	13	6	-2	-10	-18	-25	-33	-41	-49	-56	-64	-71
35	20	12	4	-4	-12	-20	-27	-35	-43	-52	-58	-67	-75
40	19	11	3	-5	-13	-21	-29	-37	-45	-53	-60	-69	-76
45	18	10	2	-6	-14	-22	-30	-38	-46	-54	-62	-70	-78

A   B   C

	WIND CHILL TEMPERATURE °F	DANGER
A	ABOVE 25°F	LITTLE DANGER FOR PROPERLY CLOTHED PERSON
B	25°F / 75°F	INCREASING DANGER, FLESH MAY FREEZE
C	BELOW 75°F	GREAT DANGER, FLESH MAY FREEZE IN 30 SECONDS

© Cengage Learning 2013

FIGURE 5-13
Wind chill index.

- Nourishment
- Rest, recovery
- Medical evaluation
- Relief from climatic conditions
- Accountability while in the rehabilitation sector
- Supplies, shelter, and adequate number of people needed to operate the rehabilitation area

Procedures should define the responsibilities within the IMS for the incident commander, supervisors, and personnel. Further, the procedures should provide guidelines for location, site characteristics, site designations, and resources required. Records should be kept on all individuals entering the rehabilitation area and medical evaluations should be documented.

NOTE Remember that the medical information from a rehabilitation sector/group may be deemed confidential medical records. Check with your local legal advisor.

FIGURE 5-14
Medical gear should be staged in the rehab area.

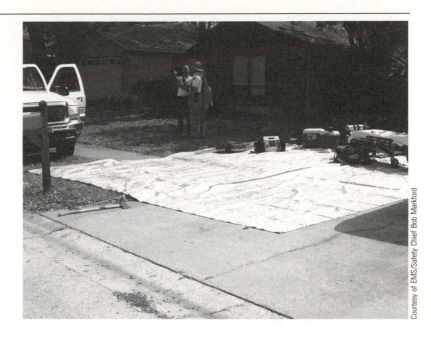

Courtesy of EMS/Safety Chief Bob Markford

SUMMARY

Fireground activities account for the highest percentages of firefighter deaths and injuries compared to any other one specific type of duty. It is essential that the safety and health program address safety issues relating to these types of incidents. This requires that the safety program manager understand operational priorities and their application to personnel safety. The operational priorities are always life safety (including the responders), incident stabilization, and property conservation. Understanding the priorities helps to perform risk assessment, which should be based on the premise that we will risk much to save much, we will risk little to save little, and we will risk nothing to save what is already lost.

Hazards faced by responders at fire incidents vary by the type and complexity of the incident. To minimize exposure to the hazards, the safety program must include components relating to personal protective equipment, require the use of an incident management system, provide for an accountability system, ensure that rapid intervention companies are deployed, and have a rehabilitation process. Personal protective equipment concerns are based on design and purchase, use, care, maintenance, and inspections.

The type of incident management system may vary provided that it meets certain criteria. Within the incident management system, there must be a safety officer with the proper authority and responsibility for incident scene safety. Responders and incident commanders have the responsibility to understand the concepts of air management when crews are relying on SCBA in a hazardous environment. Accountability systems provide for a strict accounting of all personnel throughout the incident. When an unexpected situation occurs, the incident commander must have a team available to respond and assist. This role belongs to the rapid intervention company. Prolonged physical and mental stress can adversely affect emergency responders at the incident scene. Proper procedures for rehabilitation must be part of the incident management plan.

Concluding Thought: Having a practical system in place to handle fire-based incidents, including consideration for personnel and equipment, and having the processes in place for safety mitigation will result in safer fireground operations.

REVIEW QUESTIONS

1. Property conservation is the first incident priority.
   a. True
   b. False
2. Explain how the first incident priority can have a positive impact on firefighter safety.
3. List three hazards commonly found at building fires.
4. Which of the following is not an incident priority?
   a. Preplanning
   b. Incident stabilization
   c. Property conservation
   d. Life safety
5. List the ten firefighting orders as related to wildland fires.
6. List the components of an effective PPE ensemble for firefighting.
7. List five characteristics of an effective accountability system.
8. List the positions in the command staff and general staff of an IMS.
9. What three authorities must an incident safety officer be given to be effective on the fireground?
10. How often should firefighting personnel inspect their PPE?
    a. Annually
    b. Weekly
    c. Monthly
    d. Daily and after each use

ACTIVITIES

1. Review your department's PPE inspection program. What changes do you feel are necessary to make it more effective?

2. Review your department's procedures for the assignment of rapid intervention companies. Do you use them? Are they effective? What improvements can be made?

3. After performing the risk identification process described in Chapter 3, make a matrix for your jurisdiction relating to risks found at building fires.

4. Review your department's rehabilitation SOP. Does it meet the requirements discussed in the chapter?

5. Discuss air management with your crew. Hold a drill to determine the average air consumption of the crew members.

NOTES

1. Although the SCBA unit is usually placed on the apparatus, many departments supply personal face pieces.

CHAPTER **6**

Safety at the Medical or Rescue Emergency

LEARNING OBJECTIVES

Upon completion of this chapter, you should be able to:

- Discuss the hazards faced by responders at emergency medical or rescue incidents.
- Explain methods of minimizing and preventing injuries associated with hazards found at emergency medical incidents.
- List various requirements and uses of commonly used personal protective equipment.
- Explain the requirements for infection control.
- Discuss procedures that can be used to meet the requirements of infection control.
- Discuss the systems used for scene accountability and incident management.

In February 2010, a fire police captain was fatally injured when he was struck by a motor vehicle while positioned at a controlled intersection. The victim was dispatched to the scene of a motor vehicle incident to control traffic. He had placed five warning devices (lime green traffic cones) across the two-lane roadway and had lighted a flare adjacent to the traffic cones. A short time later, as he was standing near the middle of the blocked-off roadway with his back to oncoming traffic, a motor vehicle ran through and over the cones, striking him. After initial first aid was administered by two fire fighters, the victim was transported to the local hospital by ambulance, where he was pronounced dead about 2½ hours later. Factors found to contribute to the death included:

- No advance warning to motorists of the blocked-off roadway.
- The inconspicuousness of the victim.
- The victim had his back to oncoming traffic.

Recommendations:

- Ensure that the placement of warning devices (portable signs, traffic cones, flares, and portable changeable message signs) informs drivers of what to expect when approaching an incident scene.
- Ensure that personnel controlling traffic wear high-visibility apparel and helmets.
- Ensure that standard operating guidelines include guidance on identifying and maintaining a safe location while working in or near moving traffic.
- Ensure that a personnel accountability system is in place and adhered to during emergency operations.
- Utilize state and local departments of transportation for additional resources.
- Consider participating in the establishment of local traffic incident management committees.

As responses to vehicle incidents will in almost every case put the responders at risk of being injured or killed by other vehicles, effective traffic and scene management is required.

Source: Information for this case study was taken in part from CDC/ NIOSH Fire Fighter Fatality Investigation Report F2010-06. http://www .cdc.gov/niosh/fire/reports/face201006.html

INTRODUCTION

Whereas Chapter 5 focused on fire scene and incident safety, this chapter focuses on the types of incidents that most responders face most of the time. It is not uncommon to find rural, suburban, and urban departments involved in EMS responding to EMS incidents 80% or more of the time. Therefore EMS safety is a critical and integral part of any department's safety and health program.

Information presented in this chapter relates to the following Firefighter Life Safety Initiatives:

11. *National standards for emergency response policies and procedures should be developed and championed.*

12. *National protocols for response to violent incidents should be developed and championed.*

HAZARDS FACED BY RESPONDERS

What makes the emergency medical incident scene so dangerous? There are many factors to consider when answering this question. One answer might be that, like fire officers, EMS crews must make any number of critical, time-sensitive decisions while on EMS alarms, such as the following:

- The safest and quickest route to the scene
- Drug dosages
- Treatment regimes
- The most appropriate form of transportation and the receiving facility
- Choosing the best interpersonal approach to the patient

Obviously, the decisions occupy a great deal of the responders' thought processes; however, the responders must factor in safety in all of these decisions. A good example might be the response to a multiple car accident on a divided highway that runs from north to south. The accident may be in the northbound lanes, but the ambulance is approaching from the southbound lanes. The quickest means of reaching the patient would be to drop off a responder and have him or her cross the traffic to reach the victim while the driver went to the next exit and came around. The safest approach (considering, of course, the distance to the exit) would be for both responders to continue to the exit and turn around. This might add an additional few minutes to arrival on the scene, which must be weighed against the threat to the responder in crossing the road. Which approach would you take?

Another question is the validity or accuracy of information from dispatch. Have you ever heard of a response to a "man down" that turned out to be a shooting? Not usually the dispatcher's fault, this breakdown of information commonly occurs at the calling party's end. With the increase in cellular phone use, alarms are received sooner, but often the caller is a passerby and does not know a lot about the emergency.

The responders also face safety issues when dealing with the patients, family, and bystanders. Remember that EMS events, as with all emergency incidents, are emotional events; people at the scene may not be thinking rationally and they expect immediate action from the responders. Try to tell the frantic father of a child who has been electrocuted about the safety issues and the need to cut the power. Or try to tell the outraged family member of the shooting victim that you must wait for the police until you can enter the house.

NOTE Many alarms are received by cell phones; often the caller is a passerby, however, and does not know a lot about the emergency or the location.

As in some fire-related incidents, responders to routine emergency medical calls often develop a complacence toward the type of calls (see **Figure 6-1**). How many chest pains calls have you responded to? Do you approach the patient with less caution now than you did your first days on the job? How many helicopter landings? Do you take the same precautions now as you did for your first landing zone setup? The answer very well may be yes, and that is the correct approach. However, the reality is that because you have attended numerous similar incidents and everything has gone well, you may let your guard down. This chapter reminds, reinforces, and, in some cases, gives a new look at EMS incidents and the threats to responders.

FIGURE 6-1
EMS incidents are a safety concern for responders regardless of how routine the incident may seem.

© Cengage Learning 2013

General Hazards for All EMS Incidents

Once the responder has arrived on the scene, the incident can be viewed in four phases:

1. Gaining access
2. Approaching and first contact with the patient
3. Providing care
4. Packaging and placing the patient in the transport unit

There are obviously additional issues associated with a medical incident, such as response to the scene, but they are addressed elsewhere in this text.

The first phase of the incident is to gain access to the location, whether it be a home, apartment, hotel, restaurant, or vehicle. Upon arriving at the incident scene, park and secure your vehicle in a safe, but accessible, location.

One consideration concerning gaining access are those situations in which forced entry is required. There are a number of concerns, not the least of which is security after the patient is removed. Forcible entry is necessary when a patient is unable to come to the door. Law enforcement should be requested to respond to all forcible entry situations and, where possible, the forcible entry should wait until their arrival. Forcible entry is in itself a safety hazard. By definition, forcible entry requires that something will be broken, such as glass or door frames. Responders should have on hand the necessary PPE and the proper tools when attempting forcible entry. If the equipment and tools are not available, then additional resources should be called. Before you force entry, consider that another opening may be unlocked. Many times a second-floor window or balcony door may be left open or unlocked, requiring only the use of a ladder.

NOTE Before any forced entry occurs, make sure you are at the right location.

In most cases, forced entry is not required and the responder can enter the location as any other person would. But there are still considerations about how you enter and what you see when you do enter. The responder should always carry a flashlight at night and consider carrying it during the day at some locations. In the event the patient is in a basement or other dimly lit area, the flashlight will be necessary, so it is best to carry it always. When looking around on entry, the same considerations should be noted as on arrival outside. Look for access to the patient and consider transport options, poorly repaired stairs, rugs, tripping hazards, and animals. An animal present in the house should prompt the responder to ask that the animal be secured in another room. Although the animal may seem friendly or small enough, most animals are very protective of their owners and an attempt to touch or move an owner may induce an unwanted response from the animal.

Once access has been made and the patient located, phase two, approach and first contact, begins (see **Figure 6-2**). There are a number of things to consider when first contact is made, most of which deal with interpersonal skills. An approach to a conscious

FIGURE 6-2

EMS responders should be alert for safety issues when first contact is made with a patient.

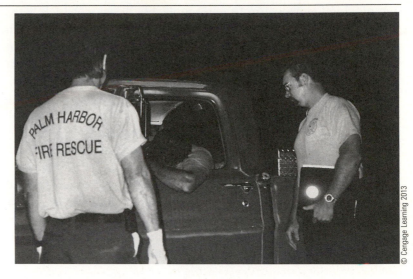

© Cengage Learning 2013

patient should start with an introduction and why you are there. If hostility toward the responder exists, it will be apparent at this time. Be aware of threats or statements that indicate that the patient may not want your help. Again, this may be an indication that police assistance is needed and a retreat necessary. Note the patient's hands and surroundings and make sure nothing is present that might be used to hurt you or cause injury as you work around the patient. Glass on the floor is a good example, considering you will probably kneel down to take vital signs. Always approach cautiously and move items out of the way that may pose problems in the next phases of the incident.

Unconscious patients should be approached with caution as well. Is it possible that you might be waking up a patient who is just under the influence of drugs or alcohol and passed out? Or is it possible that the patient is just sleeping and could react violently when awakened? Again, a cautious approach and being prepared to back off is required.

NOTE During first contact, the responder must use a variety of interpersonal skills.

Tone of voice is important: Commanding respect would be necessary in a nightclub, while soft and gentle might be in order for a small child at home. An understanding of personal space is important, remembering that in the space in which the responder works, taking vital signs, holding spinal support, and the like involve touching the patient. It is generally accepted in the United States that intimate space is 18 inches around a person. Therefore, any unfamiliar responder immediately invades this space and the patient may experience a feeling of uneasiness, leading to an unwanted reaction. It is, therefore, very important for the responder to develop trust early on.

After, and almost simultaneous with, first contact is the providing of care phase of the response. In this phase the responder is again faced with several safety challenges, one of which is exposure to communicable diseases, which is discussed under Infection Control later in the chapter. Others include injury while moving the patient, needle-sticks, and contact with objects around the patient. Usually if the necessary precautions have been taken in the first two phases and the situation does not involve communicable diseases or hostility, this phase can proceed relatively safely.

The final phase of the EMS incident to be discussed in this chapter is packaging and placing the patient in the transport unit. Now is the time when many of the mental notes that were made on arrival become important. Are the stairs safe to carry the patient? If not, what other means are available? Consider that stairs not only have to support the weight of the responders, but also the patient being carried. Perhaps a fire department ladder truck with a bucket might be a safer alternative. Were there any obstructions or tripping hazards, either inside or outside, on the approach? Are rugs and carpets secured down? If not, roll them up to create a clear path. Finally, are enough people present to make the move safely, considering outside weather conditions, the patient's weight and location, and available equipment?

Hazards Associated with Vehicle Accidents and Incidents on Roadways

As compared to incidents inside a structure, operations associated with incidents on roadways pose distinct safety hazards. Responders to incidents on roadways have a great deal less control over their surroundings (see **Figure 6-3**). Traffic, weather, lighting

FIGURE 6-3

For the safety of the responders, incident commanders must consider stopping all traffic on the roadway while fire and EMS operations are underway.

Courtesy of EMS/Safety Chief Bob Markford

conditions, and crowd control are concerns. Responders' safety on the highways has received a great deal of attention, and rightly so.

On arrival at this type of scene, the first order of tactical concern is where to safely locate the response vehicle. The first-arriving unit must be placed to have an initial blocking of oncoming traffic in an effort to create a safe work area. The unit needs to be parked in a manner that will provide the best level of protection for the responders. Other considerations are which location will best facilitate patient care and transport, which location will provide the best departure, and whether other emergency vehicles will be arriving. Of course, safety is the primary concern when considering vehicle placement. Larger vehicles, such as fire apparatus, should be positioned to block the scene; smaller vehicles such as ambulances and law enforcement vehicles should be placed inside the protection of the larger vehicles. Locating the unit also depends on other hazards that could be present at the scene, such as fuel spillage, downed power lines, or hazardous cargo. Always avoid parking the vehicle in a manner that requires emergency responders to cross traffic to get to and from the scene.

Traffic always presents hazards to responders to roadway incidents. Several safety principles should be applied. Create a buffer zone by using an emergency vehicle as described in the previous paragraph. Patient loading areas must be within the protected area afforded by the larger apparatus. Do not be afraid to close the road completely to protect responders' safety. This can be accomplished by using orange traffic cones and law enforcement personnel. Finally, wear protective clothing based on ANSI standards and OSHA requirements, or a vest with reflective trim regardless of whether the incident is in the daytime or after dark (See **Figure 6-4**). **Text Box 6-1** provides some general guidelines for personnel operating on highways.

FIGURE 6-4
Responders must wear protective clothing or a vest with reflective trim regardless of whether the incident is in the daytime or after dark.

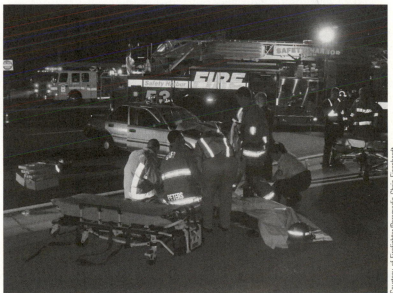

Courtesy of Firefighter/Paramedic Chris Eisenhardt

TEXT BOX 6-1

GENERAL SAFETY PRINCIPLES FOR RESPONDERS OPERATING AT HIGHWAY INCIDENTS

- Wear proper PPE including reflective vest and helmet.
- Maintain personal situational awareness, particularly of traffic, at all times.
- When personnel permits, have a spotter watching oncoming traffic at all times.
- Have a system to alert responders on the scene of a threat. This system might be an air horn on the fire apparatus similar to a fire scene evacuation signal.
- Avoid turning your back to oncoming traffic.
- Check traffic before you move.
- Look before exiting the response vehicle.
- If department policy, deploy cones and, at night, flares. Face traffic when deploying these devices. Be aware of flare use or smoking by bystanders around leaking fluids from the vehicles.

© Cengage Learning 2013

Other hazards related to roadway incidents, specifically vehicle accidents, are those hazards associated with vehicle rescue or extrication. **Vehicle rescue** is comprised of actions to gain access to, disentangle, and remove persons trapped in vehicles. Most often this is accomplished using heavy hydraulic equipment for cutting and spreading metal and car components (see **Figure 6-5**). Aside from the hazards of the accident itself, using this equipment produces the following additional hazards:

- Flying glass and metal
- Potential of rescuers' fingers and hands being caught between the vehicle and the cutting or spreading tool
- Fire hazards as a result of using powered tools
- Injuries resulting from lifting heavy tools and equipment

Any time vehicle rescue operations are being performed, an IMS should be established, a safety officer should be assigned, full protective clothing designed for thermal and abrasion resistance should be worn, and fire suppression capabilities should be present (see **Figure 6-6**).

---

vehicle rescue   Comprised of action to gain access, disentangle, and remove persons trapped in vehicles.

FIGURE 6-5
Examples of heavy
extrication equipment.

© Cengage Learning 2013

FIGURE 6-6
A vehicle extrication in
progress.

Courtesy of Palm Harbor Fire Rescue

In addition to vehicle accidents, other incidents occur on roadways that require EMS. Injuries from violence—including shootings, stabbings, and assaults—are too prevalent in the country today. Responders to these incidents should be aware of these situations and rely on law enforcement support and scene control. The hazards associated with these incidents are obvious. At a drive-by shooting, for example, who is to say the shooters will not drive by again after responders have arrived.

Environmental hazards are also a concern in roadway situations. Extreme cold or heat, rain, snow, sleet, and the duration of the operation can impact the responders' safety. Rehabilitation considerations similar to those presented in Chapter 5 should be a part of the response plan.

Chapter 4 discussed the importance of interagency preplanning and interaction. This is very important for roadway incidents. Law enforcement is needed for crowd and traffic control and to ensure the securing of the scene following incidents involving violence. Preplanning the necessary equipment, resources, and expectations of both agencies is essential to maximize safety on the scene.

INFECTION CONTROL

Infection control is a requirement for emergency responders to EMS incidents. Chapter 2 presented federal regulations and an NFPA standard that govern these programs. Specifically OSHA 1910.1030 deals with bloodborne pathogens and *NFPA 1581* gives the requirements for a fire department infection control program. Infection control is essential, not only for responders' safety, but also to maximize the safety of the patients they treat. The safety program manager may or may not be the department's infection control officer. Regardless of who it is, the law requires the organization to assign an infection control officer to handle exposures, follow-up care, and record keeping. The organization must also adopt an infection control plan to prevent the transmission of diseases.

The infection control plan is based on engineering controls, personal protective equipment, and education. Engineering controls can be based on risk management principles. For example, the risk of needlesticks can be reduced or eliminated by changing the practice of resheathing needles to just placing them in a sharps container. Personal protective clothing is discussed in the next section. Education requires training employees in the types of infectious diseases, routes of transmission, and prevention strategies.

NOTE Responders are exposed to a number of diseases—some minor, some serious, some even fatal.

In 2009, many organizations faced the threat of the H1N1 flu pandemic. This required planning not only for an increased number of patients but also for possible mass illness among responders, emphasizing the need for a strong infection control program as a part of the overall safety and health program.

An exposure by itself is not the single mechanism needed to acquire the illness; several factors must interplay for an infection to occur. As with injury prevention, the infection can be prevented by interrupting the disease process at any of these points:

Dose: the number of live organisms present

Virulence: the strength of the infecting organism

Host Resistance: the ability of the host to resist the effects of the infectious organisms

Route of Exposure: airborne, bloodborne, or foodborne

Means of Transmission: a way for the organism to gain entry into the host— injection, inhalation, absorption, or ingestion

An infection control manual can provide information regarding the listed factors for specific diseases. One of the best ways to intervene or interrupt the disease process is to prevent the means of transmission. Diseases are transmitted via direct or indirect contact, an intermediate host, or other vehicles.

Direct contact transmission occurs when physical contact occurs between the responders and the infected person. Indirect contact occurs when an infected person contacts an object and then the responder contacts the same object. Of course, in the case of bloodborne disease, there would have to be blood on the object. Another means of indirect contact is droplet contact, which occurs when an infected person coughs or sneezes and airborne droplets are produced. These microscopic droplets can travel a great distance.

Examples of transmission via an intermediate host would be diseases spread by ticks, mosquitoes, flies, or fleas. Disease spread through water and food are examples of transmission via other vehicles.

NOTE As described in Chapter 4 in the section on employee wellness and fitness, responders should take advantage of and be provided with boosters and vaccinations for diseases when available.

Responders should also be trained in personal hygiene and equipment cleanup procedures. The response station must have clearly marked areas for cleaning contaminated equipment. In terms of personal hygiene, hand washing is particularly important. Since hand-washing facilities are not always available, waterless gels that are disinfectants should be carried on the emergency vehicle. Procedures should be in place for the responders to ensure that contamination stays out of the response vehicle and then out of the station.

PERSONAL PROTECTIVE EQUIPMENT

PPE for responders to EMS incidents is specialized and differs from that of firefighting PPE. EMS responders who are cross trained as firefighters are issued firefighting PPE that is appropriate for most EMS incidents, such as vehicle rescue, because firefighting PPE provides thermal and abrasion resistance. Because firefighting PPE was described in Chapter 5, this section focuses on PPE particular to EMS emergencies that are unrelated to entrapment or rescue.

PPE requirements for **body substance isolation (BSI)** are defined in the respective regulations and standards. The EMS providers often must make a decision as to what PPE is necessary at particular incidents. **Universal precautions** should always be used any time there is a chance to come into contact with potentially infected body fluids or materials with body fluids on them. Procedures should be developed to give guidance in PPE selection for tasks at the emergency scene. **Table 6-1** is an

body substance isolation (BSI) An infection control strategy that considers all body substances potentially infectious and isolates all body substances that might be infectious.

universal precautions Term used to describe the practice of treating all patients as if they carried an infectious disease, in terms of body fluid contamination.

TABLE 6-1 PPE Matrix

TASK OR ACTIVITY	DISPOSABLE GLOVES	GOWN	MASK	PROTECTIVE EYEWEAR
Bleeding control with spurting blood	X	X	X	X
Bleeding control with minimal bleeding	X			
Emergency childbirth	X	X	X	X
Blood drawing	X			X
Starting an intravenous (IV) line	X			X
Endotracheal intubation, esophageal obturator use	X		X	X
Oral/nasal suctioning, manually cleaning airway	X		X	X
Handling and cleaning instruments with possible microbial contamination	X	X	X	X
Measuring blood pressure	X			
Giving an injection	X			
Measuring temperature	X			
Cleaning back of ambulance after a medical alarm	X	X	X	X

TABLE 6-2 Personal Protective Equipment for Body Areas

BODY AREA	PPE ITEM	REMARKS
Head	Helmet	If not provided as firefighting PPE
Head	Eye protection	To protect from objects and body fluid intrusion
Head	Hearing protection	For prolonged rescue in noisy environment
Head	Mask	To prevent airborne transmission or fluid from entering the mouth
Torso	Abrasion-resistant jacket	If not provided as firefighting PPE
Torso	Gown	To prevent fluid transmission
Arms	Protective sleeves	To prevent fluid transmission
Hands	Abrasion-resistant gloves	If not provided as firefighting PPE
Hands	Fluid-resistant gloves	If not provided as firefighting PPE
Legs	Abrasion-resistant pants	If not provided as firefighting PPE
Legs	Gown	To prevent fluid transmission
Feet	Shoes hard sole, steel toe	If not provided as firefighting PPE
Feet	Shoe covering	To resist fluid penetration
Body	Body armor	Dependent on local conditions and risks

© Cengage Learning 2013

example of a matrix that should be included as part of an infection control SOP. It is *always* the rescuer's option to do more than the minimum requires based on the circumstance. From head to toe, the EMS responder should have PPE adequate for the hazards to be faced. **Table 6-2** charts PPE for various body areas. These are recommendations and must be geared to local conditions.

EMS PPE should be cared for, disposed of, and inspected according to the manufacturer's recommendations. The inspection program should be similar to that of fire PPE; however, much of the EMS PPE is disposable. The same record keeping should be required for EMS PPE as for firefighting PPE.

INCIDENT MANAGEMENT SYSTEMS

Like fire incidents, emergency medical/rescue incidents must be managed with an IMS. This is not to say that responders must break out the command board for every heart attack call, but in reality IMS does occur on each incident. It is just that the senior EMS responder assumes many of the roles in the command staff. On multiple unit or multiple jurisdictional response, IMS is as much required on the EMS incident as it is on the fire or hazardous materials scene.

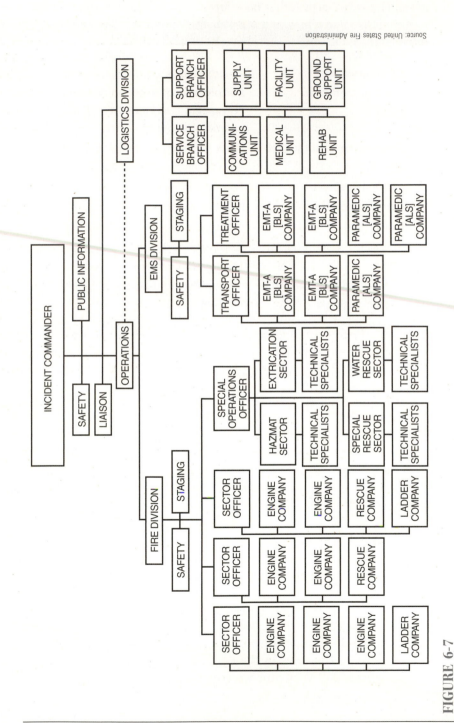

FIGURE 6-7

An example of an expanded IMS including EMS functions. Note: The EMS division is operational; the medical unit is logistical.

For an EMS incident, the IMS can be adapted. The three command staff and four general staff functions remain the same. Responsibilities within the command structure only differ depending on the type of incident. At a multicasualty bus accident without entrapment, there may be a division or group for triage, treatment, and transport. At a single vehicle in a pole with entrapment, there may be groups or sectors for hazard control, extrication, and treatment. The system is adaptable to any emergency. One common factor is the Incident Safety Officer. Although very dependent on the situation, the safety officer should be assigned any time responders are performing hazardous operations such as extrication.

NOTE On a routine medical incident the senior person or officer must assume all of the command functions, including safety.

In this role, the safety officer can ensure and require that proper PPE is being worn for the particular type of incident. **Figure 6-7** details an expanded IMS at a medical/rescue incident.

As with fire incidents, an accountability system for EMS responders must be adhered to. In some jurisdictions, some parts of EMS response are handled by private or third-service public agencies. When these agencies respond and are required to work at the incident, they must be part of and work within the IMS structure and the accountability system. The safety program manager for the EMS provider should ensure that all EMS personnel have the proper training and equipment to work within these two systems. This also can be part of interagency coordination.

SUMMARY

For most jurisdictions, EMS emergencies comprise the highest percentage of incidents. The hazards faced by responders are increased by uncertainties and the need to interact closely with the patients. Medical or rescue incidents can be generally divided into those that occur inside and those that occur outside. Inside structure incidents, often viewed as routine, can cause injuries. A step-by-step approach and an awareness of surroundings can reduce this potential.

Outside incidents usually involve vehicle accidents. Hazards such as the vehicle, the fuel for the vehicle, traffic, and the environment cause additional problems for the responders. Anticipating these hazards through training on proper vehicle placement and control of the scene can reduce exposure to these hazards. Utilizing law enforcement for crowd and traffic control is also required.

Infection control is an additional concern in medical response. Through the use of engineering, education, and PPE, the risk of acquiring an infectious disease can be reduced. The use of PPE is both required and an excellent means to interrupt the disease transmission process.

The IMS must be adaptable to EMS events and provide for the same command functions and structure as that of fires. The system must include a method for scene accountability of EMS responders, particularly if the EMS responders are from a different agency.

Concluding Thought: EMS and fire events have many of the same requirements for success. Engineering, education, and PPE are three requirements essential to the safety program.

REVIEW QUESTIONS

1. List four hazards associated with EMS emergencies that occur indoors.
2. List three hazards that would be present outside on a highway at a vehicle accident.
3. Which of the following practices should a responder to a vehicle accident *not* take?
   a. Create a buffer zone
   b. Create a nontraffic lane
   c. Park so that traffic must be crossed to access patients
   d. Wear reflective vests
4. What three functions might be in the IMS for a multicasualty incident?
5. List the five factors that must interplay for an infection to occur.
6. Which of the factors in Question 5 is easiest to control?
7. In terms of the spread of infection, compare the terms *direct contact* and *indirect contact.*
8. What are the three components involved in vehicle rescue?
9. What EMS PPE components might also be part of firefighting PPE?
10. List five tasks or activities that should be included in a required PPE matrix.

ACTIVITIES

1. Review your department's procedures and practices for responding to vehicle accidents. Do they apply the safety principles as discussed in this chapter?
2. Review your department's infection control plan and compare it to the federal requirements. Do you have an infection control officer, and does this officer regularly respond to incidents? Does the plan provide for engineering controls, education, and appropriate levels of PPE?

CHAPTER 7

Safety at Specialized Incidents

LEARNING OBJECTIVES

Upon completion of this chapter, you should be able to:

- Describe the safety issues related to hazardous materials incident responses.
- Describe the safety issues relating to technical rescue operations.
- Explain safe procedures to be used during helicopter landing zone operations.
- List specific safety issues relating to operations at civil disturbances.
- List specific safety concerns when responding to terrorism events.
- List specific safety concerns when responding to natural disasters.

On March 22, 2006, a captain died from complications of a drowning incident that had occurred two days prior. The victim was in a training class with another student to become certified on a personal watercraft/waverunner (PWC) to conduct surf rescue operations for his jurisdiction. While the victim participated in training evolutions as a passenger on the back of the PWC, he fell off and re-boarded numerous times while the engine was running. After a wave knocked both trainees from the PWC, the victim reported that he was too tired to re-board and seemed confused. The instructor attempted to drag the captain into shore using an 18-inch tow strap attached to the rear of the PWC, but this was very difficult due to the surf conditions and the engine's water propulsion hitting the victim's upper body. After numerous attempts to hang on, the captain became lost in the surf. A few minutes later, the instructor found the captain and swam him to shore. The victim was retrieved from the surf and advanced life-saving efforts were initiated until the ambulance arrived on the scene and transported him to a local hospital. He was airlifted to a regional medical center where he died two days later from the injuries received during the incident. NIOSH investigators consider the drowning episode to have been triggered by a combination of exhaustion, hypothermia (rectal temperature less than 95.7° Fahrenheit [F]), and carbon monoxide poisoning (carboxyhemoglobin level estimated to be 20%). To minimize the risk of similar occurrences, fire departments, police departments, and/or agencies with surf rescue teams should:

- Ensure that PWC are never boarded by an operator or a passenger while the engine is running.
- Enhance current standard operating procedures to include requirements for rescue boards on all PWC during emergency responses and training.
- Enhance current standard operating procedures to include requirements for issued personal protective equipment (PPE).
- Enhance current standard operating procedures for training operations to duplicate the same level of response as an actual emergency.
- Provide an emergency medical services unit at all training sessions.
- Consider conducting a job analysis of surf rescue to determine its fitness and strength requirements.

Water rescue is one of the technical rescue areas identified in the NFPA technical rescue standard. Technical rescues (along with other specialized incident response activities) require specialized training, equipment, and safety procedures specific to the related hazards. Safety officers assigned to these type incidents also must have experience and training unique to the type of emergency.

Source: Information for this case study was taken in part from CDC/NIOSH Fire Fighter Fatality Investigation Report F2006-14. http://www.cdc.gov/niosh/fire/reports/face200614.html

INTRODUCTION

The discussion of safety issues at emergency scenes would not be complete without a look at safety and special operations. These particular types of incidents include hazardous materials response, technical rescue incidents, helicopter landing zone operations, operations at civil disturbances, terrorism events, and natural disasters.

These specialized incidents are addressed in a separate chapter because some departments may not be involved in all of these operations. Some departments may not be involved in any of them. The local conditions and the services that a particular department provides dictate what safety measures and procedures must be in place. Each type of incident is discussed and basic safety concerns presented. The following Firefighter Life Safety Initiatives apply to specialized incidents:

3. *Focus greater attention on the integration of risk management with incident management at all levels, including strategic, tactical, and planning responsibilities.*

4. *All firefighters must be empowered to stop unsafe practices.*

5. *Develop and implement national standards for training, qualifications, and certification (including regular recertification) that are equally applicable to all firefighters based on the duties they are expected to perform.*

6. *Develop and implement national medical and physical fitness standards that are equally applicable to all firefighters, based on the duties they are expected to perform.*

HAZARDOUS MATERIALS INCIDENTS

Responding to a hazardous materials incident is common for most fire departments. What differs significantly is the level of response. Some departments may have full hazardous materials teams and provide complete response services, including offensive action to mitigate the incident. Other departments may respond at the first response level and provide defensive operations only, relying on a regional hazardous materials team for offensive action and mitigation. OSHA 1910.120 and EPA regulation Section 311 govern the response to hazardous materials incidents. There are also two NFPA standards and one Recommended Practice applicable to hazardous materials:

- *NFPA 471, Recommended Practice for Responding to Hazardous Materials Incidents* (Note: This Recommended Practice was withdrawn in 2007 and incorporated into *NFPA 472* and *NFPA 473.*)

- *NFPA 472, Standard for Professional Competence of Responders to Hazardous Materials Incidents*

- *NFPA 473, Standard for Competencies for EMS Personnel Responding to Hazardous Materials Incidents*

FIGURE 7-1
Hazardous materials
response and operations
produce unique safety
concerns.

© Cengage Learning 2013

**TEXT BOX
7-1**

FIVE LEVELS OF HAZMAT RESPONSE TRAINING

First Responder Awareness

First Responder Operational

Hazardous Materials Technician

Hazardous Materials Specialist

On-Scene Incident Commander

© Cengage Learning 2013

These regulations require that certain safety measures be in effect before operations can begin. These requirements are described in Chapter 2 (see **Figure 7-1**).

NOTE As part of the overall safety and health program, the safety program manager must determine what level of response the department provides and compare that level to the training level of the responders.

There are five levels of response training, as shown in **Text Box 7-1**.

Having identified the five levels of training, the program safety manager must review department procedures and practices to ensure that the responders are operating within the level in which they have been trained and within the level of equipment available. A serious incongruence would occur if the department responded routinely to hazardous materials releases and plugged the leaks, but had only been trained to a first responder level.

There are numerous safety-related issues regarding a hazardous materials incident. The risk factors increase with the level of service provided. Hazardous materials PPE is a complete subject in itself. Regulations and standards require the use of incident command systems and backup teams with the same level of protection as the entry team. However, many of these issues relate only to departments that provide hazardous materials technicians and do offensive mitigation strategies. These safety measures should be addressed in the hazardous materials team's operating procedures.

One area of hazardous materials response that is applicable to all responders is the initial response and arrival. Procedures and training for this segment of the incident should focus on safety and the initial response. One method to use to maximize the safety of the initial response is to remember the acronym RAID, whose letters stand for Recognize, Approach, Identify, and Decide. See **Text Box 7-2.**

TEXT BOX 7-2

THE RAID PROCESS

- *Recognize.* The recognition phase is when the responder recognizes that an incident might involve a hazardous substance. The recognition process begins with preplanning. Being aware through preplanning of what hazards are present in a given response area helps to determine the precautions that will be taken at an incident at that specific location. Clues to the presence of hazardous materials also present themselves at the time of dispatch: The location, occupancy, and the name of the business can provide clues. For example, a report of a person(s) down behind Jones Pool Chemicals should give the responders a pretty strong clue that hazardous materials may be involved. The presence of multiple victims with similar symptoms at the same location is also a good clue.

- *Approach.* During the approach phase, several things must be considered. If the responders suspect that a hazardous substance is involved, the approach should be cautious and, if possible, from uphill and upwind. During the approach, the responders should be looking for other clues, such as strange-colored smoke, vapor clouds, and employees running from the area. If necessary, park the response unit away from the scene until further analysis can be performed. In the Jones Pool Chemicals example, the crew should be alert for vapor clouds or smells such as chlorine fumes.

- *Identify.* Once the scene has been safely approached by responders, and if there is a strong suspicion of a hazardous substance incident, they may alert the hazardous materials response team. However, if further information is required, the first responder can start the identification process. Text Box 7-3 outlines some of the resources available for hazardous substance identification.

- *Decide.* The final step in the RAID acronym is to make a decision about any necessary mitigation measures or control actions. At this point, the responders should have a reasonably good idea whether they are equipped and trained to proceed and take action. If the hazard exceeds the capabilities of the responders, specialized teams should be called.

TEXT BOX 7-3

SOME RESOURCES AVAILABLE FOR HAZARDOUS SUBSTANCE IDENTIFICATION

- *Occupancy and Location.* The mere location of an incident or the occupancy of the building or vehicle may provide clues to identification.

- *Placards and Labels.* Hazardous substances in transit will be placarded according to U.S. Department of Transportation guidelines. However, if the load is less than 1,000 pounds, some materials are not required to have a placard. If the placard contains the four-digit identification number, the product can be identified. If not, then just a generalization as to the type of material is provided, for example, a flammable gas. Labels are similar to the placards but are found on the individual packages.

- *Shipping Papers.* Regardless of the mode of transportation, some type of shipping papers should be with the operator—in a train they are in the engine, or in a truck they are in the cab. These papers provide identification of the product as well as hazards.

- *NFPA 704 System Placard.* Fixed facilities may have an *NFPA 704* placard on the area where hazardous substances are stored or used. This placard is a diamond divided into four small diamonds. Each diamond is color coded for the hazard and assigned a number from 0 to 4, with 0 being no hazard and 4 being an extreme hazard. The color code is red for flammability, blue for health, yellow for stability, and the white diamond for special information.

- *Material Safety Data Sheets.* Fixed locations that store or use hazardous substances are required to maintain material safety data sheets, which provide product identification information, hazards, and procedures for emergencies.

- *Employees and Occupants.* Employees working at the incident location may also be able to provide product identification information. Remember, however, that employees often downplay the event. Information should be backed up with reference information. Certain facilities may also have internal fire brigades.

- *ChemTrec.* May be called for assistance in chemical emergencies not only to assist in identification but also for emergency action. ChemTrec can usually put the caller in touch with product experts or shippers. (800) 424-9300

- *Reference Books.* Although a number of reference books exist, no emergency response unit should be without—at a minimum—the Department of Transportation's *Emergency Response Guidebook*, "the Orange Book" (see Figure 7-2).

(continues)

■ *Symbols.* Symbols may provide at least general information regarding the hazards of the chemicals. Together with the specific information using the resources above regarding the physical and health hazards of the hazardous chemical, symbols can be useful in the identification process.

■ *Monitoring Equipment.* Numerous meters and other tools are available to sample and assist in product identification. However, the use of these tools is usually beyond the scope of the initial responder.

© Cengage Learning 2013

FIGURE 7-2
Placards and labels help responders identify products.

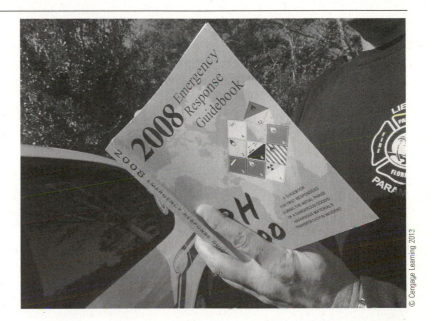

Awareness of safety issues at a hazardous substance emergency is important for all responders. Individual departments must assess their level of response and develop appropriate safety procedures.

TECHNICAL RESCUES

NFPA 1670 is a standard that addresses technical rescue operations and includes the following categories of technical rescues:

■ Rope Rescue
■ Structural Collapse Search and Rescue
■ Confined Space Search and Rescue

- Vehicle Search and Rescue
- Water Search and Rescue
- Wilderness Search and Rescue
- Trench and Excavation Search and Rescue
- Machinery Search and Rescue
- Cave Search and Rescue
- Mine and Tunnel Search and Rescue
- Helicopter Search and Rescue

Each of these types of incidents requires a great deal of expertise and specialized equipment to handle. Much like the hazardous material responders' training levels, the NFPA has developed training levels for technical rescue responders. These training levels are described in **Text Box 7-4**.

These types of incidents present a number of safety issues to the responder. The safety program manager along with the organization's management must determine the level of response to be provided and equip and train personnel accordingly (see **Figure 7-3**).

TEXT BOX 7-4

TECHNICAL RESCUE RESPONDER TRAINING LEVELS AS DEFINED IN *NFPA 1670*

- *Awareness.* The awareness level represents the minimum capability of responders who as part of their normal job duties could respond to, or be the first on the scene of, a technical rescue incident. This level can involve search, rescue, and recovery operations. Members of a team at this level are generally not considered rescuers.

- *Operations.* The operations level represents the capability of the responder to perform hazard recognition, use equipment properly, and perform the techniques necessary to safely and effectively support and participate in a technical rescue incident. This level can involve search, rescue, and recovery operations, but usually operations are carried out under the supervision of technician-level personnel.

- *Technician.* The technician level represents the capability of the responder for hazard recognition, proper equipment use, and performance of techniques necessary to safely and effectively coordinate, perform, and supervise a technical rescue incident. This level can involve search, rescue, and recovery operations.

FIGURE 7-3
Technical rescue operations require specialized training and equipment.

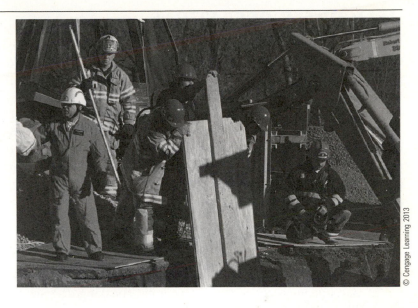

© Cengage Learning 2013

Each technical rescue response presents different safety issues ranging from air supply in a confined space to the threat of a secondary collapse in a structural collapse rescue. Generally, technical rescue incidents last longer than fire and emergency medical alarms.

NOTE Rehabilitation is of prime concern during these incidents.

Technical rescue incidents must be managed with an IMS and a safety officer in place.

The safety officer should be familiar with the operation in question. General safety issues relating to technical rescue incidents include:

- Environmental conditions. Responders will be exposed for long periods of time.
- Stability of the building, trench, or confined space.
- Duration of available air supply, if necessary.
- Resources of personnel and equipment.
- Safe atmosphere—Flammable gases? Oxygen deficient?
- Rehabilitation of responders.
- Backup teams similarly trained and equipped.
- Logistics, food, rest, and so forth for long-term operations.

Each technical rescue incident must be evaluated for risk versus gain. Often these incidents are actually body recovery situations; responders' safety must not be compromised. Suggestions for the incident safety officer assigned to a technical rescue incident are provided in Chapter 9.

HELICOPTER OPERATIONS

Emergency responders are often called on to perform operations dealing with helicopters. These operations may include working with the helicopter during a rescue, deploying responders from the helicopter into water, setting up a landing zone for a medivac helicopter, or providing ground support for a helicopter fighting a wildfire. Clearly, some of the incidents are specialized and require a great deal of training and coordination with the helicopter crew; however, some general safety precautions can be taken whenever operating around helicopters. Primarily this section deals with ground support operations and approaching the aircraft. The most common ground support operation involving helicopters is the establishment of landing zones. Several safety principles present during this function.

Landing Zone

Be sure to select a proper landing zone. This zone will differ with jurisdiction and between day and night operations. Generally speaking, the size of the landing zone depends on the type and size of the aircraft. Landing zones can range from 60 to 120 feet in length and width. The landing zone should be on as level ground as possible. A pitch of greater than 8 degrees to 10 degrees is risky.

NOTE The landing zone should be free from flying debris. Remember, if the aircraft weighs 2,000 pounds, then 2,000 pounds of air must be displaced for the aircraft to fly.

This produces very high wind on landing and takeoff. Garbage can lids and other debris can be picked up and moved about in this wind, possibly even into the blades.

Marking the landing zone is based on local procedures. Many jurisdictions require five markers: one on each corner of the landing area and the fifth to be placed on the windward leg in the center to show wind direction. At night, the landing zone markers must be lit. Usually the marker lights are blue. Some jurisdictions mark the landing zone with headlights of vehicles. Although this practice is acceptable, it must be cautioned that the lights must stay aimed so as to not shine up into the pilot's eyes. It is not good practice to use flares or cones.

Fire suppression must also be available at the landing zone. The fire suppression unit should be specifically assigned to landing zone duties and an IMS should be in place. The crew should have full protective clothing on and be prepared to deploy a hose line for fire suppression. The landing zone fire suppression unit should be parked so as to not be a part of an emergency should one occur.

NOTE Responders standing by or working near the helicopter should be provided with eye and hearing protection in addition to firefighting PPE.

FIGURE 7-4
Typical landing zone configuration for night operations.

Figure 7-4 shows a typical landing zone configuration for a night operation.

Crowd Control

The landing of a helicopter often draws many onlookers. Vehicle and crowd control are an important function around the landing zone. Often a helicopter will land and keep the motors running so there is no delay in departure. Nonessential people, including by-standers and responders, should be kept a safe distance from the aircraft. Although this crowd and traffic control is, in most cases, a police function, other emergency respond-ers may find themselves involved. It is a good idea to keep bystanders and nonessential emergency personnel at least 200 feet from the aircraft.

Approaching the Aircraft

In many cases emergency responders will, at some point, have to approach the aircraft, whether it be to load a patient or hook up a hose line to refill a drop tank. The safety prin-ciple to be applied in this case is to *never* approach the aircraft without permission from the pilot. After receiving permission from the pilot, the responder may approach the air-craft, but should always approach from within the pilot's view and always from the down-hill side if on a grade. **Figure 7-5** depicts safe and unsafe approach areas to helicopters during ground operations. Should a rescue basket be used, static electricity is a safety concern. The basket should touch the ground before a rescuer on the ground touches it.

FIGURE 7-5
Approach helicopters only from
the safe zone.

TAIL ROTOR

DANGER

CAUTION

SAFE

MAIN ROTOR

© Cengage Learning 2013

CIVIL DISTURBANCES

There have been a number of **civil disturbances** in the United States over the years. Emergency service responders are often on the front lines during these events, either for the treatment of injured or for fire suppression. These incidents require a great deal of coordination and cooperation with law enforcement.

Emergency response personnel operating during civil unrest are exposed to many hazards, including gunfire, being assaulted with objects that have been thrown, and increased hazards at fires due to the use of accelerants. Personnel may minimize some of these risks through prioritization of incidents, police protection, and additional PPE.

Incidents should be prioritized according to the potential risks involved. If a life threat is not present, the risk to personnel should be closely examined before the responders become the life safety concern. If the incident is a priority incident and units must respond, the police should either escort the units or ensure the area is clear prior to fire or EMS response. Emergency response units may be put into task forces in order to respond in higher numbers. This arrangement might be two engines, a ladder, a medic unit, and a battalion chief, with police escort. In the case of priority building fires, units should not commit to long operations, but instead hit fires with booster

civil disturbances Uprisings of civilians that often lead to hostile acts against law enforcement and emergency responders.

tanks and master streams. This tactic allows for a fast retreat, should the situation dictate.

NOTE Police protection is a critical component of the operational plan at civil disturbances. This protection should be planned before an event.

The police often have intelligence as to where problem areas may be and what might be expected. This information should be communicated to the fire and EMS agencies. Communications and a unified command system will ensure that responders are working under the same plan and all agencies have the same information.

Some agencies that do not normally wear body armor may elect to issue the body armor for use during civil unrest. Although SCBA would provide the necessary protection, emergency responders may wish to wear masks to limit their exposure to tear gas. These masks are lighter than SCBA and provide protection from the irritation of tear gas.

TERRORISM EVENTS

The events in New York and Virginia that shocked the nation on September 11, 2001, emphasized that firefighters, EMS responders, and law enforcement personnel are the nation's first responders for terrorism events. Emergency responders were aware of this duty long before these events, because they responded during the 1990s to explosions that shook the World Trade Center and the Federal Building in Oklahoma City. In an Atlanta bombing, a secondary device detonated after emergency responders were at the scene. Whether the secondary device was intended for responders is unknown, but the potential exists that it was. No other type response requires a more coordinated effort between agencies. As described in Chapter 4, preevent interagency planning is a necessity, including agreements on communication and the incident management (command) structure.

Terrorism can occur anywhere at any time, and an organization's health and safety program must include considerations for health and safety in responding to these events. Terrorism events are extremely complex—therefore a complete study of safety at such events is well beyond the scope of this book; however, some general safety considerations and information are provided.

Terrorism acts, using weapons of mass destruction (WMD), can be classified into five general categories. Generally, terrorism acts are chemical, biological, radiological, nuclear, or explosive (CBRNE) in nature. Emergency responders must, at a minimum, be provided with awareness training for terrorism events. Unlike a fire or vehicle crash in which the problem is evident on the arrival of responders, WMD events may not present so obviously. Biological and radiological events require measurement of air quality very early in the incident; therefore the responders should be equipped with meters to determine air quality. Many WMD events will require the same safety precautions that one would take when responding to a hazardous materials incident. Therefore, proper PPE must be available to all responders including appropriate

TEXT BOX 7-5

CLUES THAT THE EVENT MIGHT BE WMD RELATED

- Is the response to a target hazard as identified in the risk analysis?
- Has there been a threat?
- Multiple victims—all with same signs and symptoms or injuries?
- Are responders victims?
- Are hazardous substances involved?
- Has there been an explosion?
- Has there been a secondary attack/explosion?
- Does review of dispatch information provide clues?
- Any physical signs present?
  - Debris field
  - Mass casualty/fatality with minimal or no trauma
  - Severe structural damage without an obvious cause
  - Dead animals and vegetation
  - System(s) disruptions (utilities, transportation, vapor clouds)

Source: FEMA Emergency Response to Terrorism Job Aid

respiratory protection. Responders should take a cautious approach as described earlier in this chapter, watching for clues that could indicate a terrorism event. **Text Box 7-5** lists some of these clues.

Responders should be trained and prepared for mass decontamination activities. Responders must also have antidote kits available that are for use with some nerve agents.

Responders must remember that the terrorism act may be designed to create a media event; emergency responders can easily be targets. There are a number of resources available to the safety and health program manager for training in safe operation at these terrorism events, including training from the Department of Homeland Security, FEMA, United States Fire Administration, the Department of Justice, and numerous textbooks. As first responders, the emergency service will be the first agency on scene at a terrorism event and we must ensure our safety as well as that of the public.

NATURAL DISASTERS

Emergency responders are the first responders to natural disasters, including earthquakes, hurricanes, tornadoes, floods, wildfires, severe thunderstorms, and flooding. Because emergency service responders will be going to these events, they must also be given consideration in the health and safety plan. Risk assessment of natural disasters

TABLE 7-1 Hazards Associated with Natural Disasters, Their Risks, and Safety Recommendations

HAZARD	RISKS	SAFETY RECOMMENDATIONS
Massive piles of construction and other types of debris, unstable work surfaces	Traumatic injuries (including serious fall injuries) from slips, trips, and falls or collapsing materials	1. Ensure that surfaces are as stable as possible. 2. Use alternative methods, such as ladders, to access unstable work surfaces. 3. Ensure that responders have a full array of PPE, including safety shoes with slip-resistant soles. 4. Ensure that responders use full protection equipment with lifelines tied off to suitable anchorage points whenever possible.
Excessive noise	Communication and temporary hearing loss	Use hearing protection devices.
Breathing dust containing asbestos (from pulverized insulation and fire-proofing materials) and silica (from pulverized concrete), which are toxic	Short term: irritation of eyes, nose, throat, and lungs Long term: Chronic effects may depend on the extent and the duration of exposure	1. Responders should be protected from breathing dust. 2. Respiratory protection: An N-95 or greater respiratory protection is acceptable for most activities.
Heat stress from wearing encapsulating/insulating bunker gear or doing heavy work in a hot, humid climate	Significant fluid loss that frequently progresses to clinical dehydration, raised core body temperature, impaired judgment, disorientation, fatigue, and heat stroke	1. Adjust work schedules, rotate personnel, add additional personnel to work teams. 2. Replenish fluids (1 cup water/sports drink every 20 minutes) and food (small, frequent, carbohydrate meals). 3. Monitor heart rate. If over 180 beats per minute minus age for more than a few minutes, stop work and rest immediately. 4. Provide frequent medical evaluation for symptoms and signs of heat stress, such as altered vital signs, confusion, profuse sweating, and excessive fatigue. 5. Provide shelter in shaded areas and the ability to unbutton and remove bunker gear.

(continues)

TABLE 7-1 Continued

HAZARD	RISKS	SAFETY RECOMMENDATIONS
Confined spaces (limited openings for entry and exit, unfavorable natural ventilation)	Low oxygen; toxic air contaminants; explosions; entrapment; and death by strangulation, constriction, or crushing	1. Purge, flush, or ventilate the space. 2. Monitor the space for hazardous conditions. 3. Lock out/tag out procedures for power equipment in or around the space. 4. Use appropriate PPE such as SCBA. 5. Light the area as much as possible. 6. Establish barriers to external traffic such as vehicles, pedestrians, or other hazards. 7. Use ladders or similar equipment for safe entry and exit in the space. 8. Use good communications equipment and alarm systems. 9. Have rescue equipment nearby.
Potential chemical exposures	Eye, nose, throat, upper respiratory tract, and skin irritation; central nervous system depression; fatigue; loss of coordination; memory difficulties; sleeplessness; and mental confusion	1. Fire Fighting: Use SCBA with full face piece in pressure demand or other positive pressure mode. 2. Entry into unknown concentration: Use SCBA gear. 3. Rescue operations with fumes present: Use gas mask with front-mounted organic vapor canister (OVC) or any chemical cartridge respirator with an organic vapor cartridge
Traumatic stress after horrific events	See Chapter 8 for information on critical incident stress	See Chapter 8 for information on critical incident stress management.
Electrical, overhead power lines, downed electrical wires, cables	Electrocution	1. Use appropriately grounded low-voltage equipment. 2. Interact with power company to ensure downed lines are not energized.

Hazard	Health Effects	Recommendations
Carbon monoxide risk from gasoline- or propane-powered generators or heavy machinery	Headache, dizziness, drowsiness, or nausea progressing to vomiting, loss of consciousness, and collapse, coma, or death under prolonged or high exposures	1. Use CO warning sensors when using or working around combustion sources. 2. Shut off engine immediately if symptoms of exposure appear.
Eye injuries from dust, flying debris, and blood	Bloodborne pathogen infection, eye injury	1. Use goggles or face shield and mask for those handling human remains while recovering deceased. Make sure to cover the nose and mouth to protect the skin of the face and the mucous membranes. 2. Use safety glasses with side shields as a minimum by all responders. An eyewear retainer strap is suggested. 3. Consider safety goggles for protection from fine dust particles, or for use over regular prescription eyeglasses. 4. Any worker using a welding torch for cutting needs special eyewear for protection from welding light, which can cause severe burns to the eyes and surrounding tissue. 5. Only use protective eyewear that has an ANSI Z87 mark on the lenses or frames.
Flying debris, particles, and handling a variety of sharp, jagged materials	Traumatic injuries ranging from minor injuries requiring first aid to serious, even disabling or fatal, traumatic injury	1. Use safety glasses with side shields as a minimum. 2. Consider safety goggles for protection from fine dust particles. 3. Educate responders regarding safe work procedures before beginning work. 4. Provide responders with a full array of PPE, including hard hats, safety shoes, eyeglasses, and work gloves. 5. Ensure that responders do not walk under or through areas where cranes and other heavy equipment are being used to lift objects.

(continues)

TABLE 7-1 Continued

Hazard	Risks	Safety Recommendations
Working with numerous types of heavy equipment, including cranes, bucket trucks, skid-steer loaders, etc.	Traumatic injury, including serious and fatal injuries, due to failure or improper use of equipment, or responders being struck by moving equipment	1. Train responders to operate equipment correctly and safely. 2. Ensure operators are aware of the activities around them to protect responders on foot from being struck by moving equipment. 3. Ensure that responders do not walk under or through areas where cranes and other heavy equipment are being used to lift objects.
Rescuing victims, recovering deceased, handling human remains, contact with surfaces contaminated with blood and body fluids	Blood, bloody fluids, body fluids, and tissue are potential sources of blood-borne infections from pathogens including Hepatitis B, Hepatitis C, and HIV.	Standard precautions (universal precautions) should be strictly observed regardless of time since death.

Source: Adapted from NIOSH Publication at http://www.cdc.gov/niosh/emhaz2.html#haz1

is essential for the local department. Because certain areas of the country are more prone to different types of disasters, it is necessary to develop the safety and health program around the actual threat. It is unlikely that a firefighter in Ohio would have to be prepared for a hurricane, but that Ohio firefighter would have be very prepared for tornadoes. A Florida firefighter has a good chance of responding to the aftermath of a hurricane, but less chance of responding to an earthquake.

In some ways, like the terrorism events described in this chapter, response to natural disasters can be very complex, although many of the hazards faced will be similar regardless of the type. **Table 7-1** provides an overview of the hazards associated with natural disasters, their risks, and safety recommendations.

After a natural disaster, emergency responders will be working with numerous other agencies during the response and recovery phase. For the safest possible working environment, interagency coordination with a number of agencies, including emergency management, law enforcement, public works, local utility companies, and in many cases the military or National Guard, is important. Preevent relationships should be established so that the responders understand each others' roles, responsibilities, and limitations.

Natural disaster operations are long-term and often widespread events. Concepts presented elsewhere in this textbook—incident management, accountability, and rehabilitation—must have priority in the operation.

SUMMARY

For the purpose of this section, response to specialized incidents includes responses to hazardous materials incidents, technical rescues, helicopter operations, civil disturbances, terrorism, and natural disasters. Each of these requires a specialized response and produces numerous unique safety concerns for the responders. While some of these types of incidents may not be handled by every organization, it would be hard to imagine any organization that does not have the potential to respond to some of these specialized incident emergencies. In some cases the organization may function as the first responders; in others the response may be with a specialized team. Hazardous materials operations are governed by federal regulations that require responders to be trained to a minimum level, to operate with proper PPE and in a buddy system, and to have site-specific safety plans and an IMS. Training is based on five levels of response. The safety program manager must access the responder's level of training and ensure it is commensurate with the type of operations being performed.

Technical rescue is a specialized area of response that requires specialized training and equipment. Technical rescue includes rescue from situations that are disciplines within themselves. From a safety standpoint, although some generalization regarding safety can be made, these situations require a great deal of expertise on behalf of the responder, the incident commander, and the assigned incident safety officer.

Many departments are using helicopters for fire suppression and, more likely, for medivac operations. At a minimum, emergency responders may be required to provide ground support for these operations. Critical ground support safety issues relate to the selection and protection of a landing zone, crowd control, and the approach to the aircraft. Local conditions and the operator of the helicopter dictate specifics regarding these issues.

Response and operations in areas of civil disturbances require close coordination with law enforcement, including unified command in the IMS. Through the use of prioritization of calls in the affected area, the use of alternative tactics, task force response, and increased PPE, a level of safety can be provided.

Terrorism events also require the close coordination between law enforcement and the emergency responders. Acts of terrorism may include chemical, biological, radiological, nuclear, or explosive weapons of mass destruction. Intended to create media attention, such incidents often involve mass casualties and may have secondary devices designed to injure emergency responders.

Finally, natural disasters include weather-related events, floods, wildfires, and earthquakes. Although some specific areas of the country are more likely than others to experience specific natural disaster threats, there are general hazards and risks that must be considered in the safety and health program.

Concluding Thought: Emergency responders are required to handle much more than fire and EMS incidents. The issues related to safety increase with the various other incidents to which we must respond.

REVIEW QUESTIONS

Refer back to chapter 2 for questions 1 and 2.

1. A person who normally responds to releases of hazardous substances and takes offensive action by plugging a leak should be trained to what level?
   A. First responder operations
   B. First responder awareness
   C. Hazardous materials technician
   D. Incident commander

2. A person who normally responds to releases of hazardous substances and takes no offensive action other than notification to start the response plan should be trained to what level?
   A. First responder operations
   B. First responder awareness
   C. Hazardous materials technician
   D. Incident commander

3. What does the acronym RAID stand for?
4. List and explain three pieces of information available to help the responder identify a released hazardous substance.
5. In terms of terrorism, what does the acronym CBRNE stand for?
6. What areas of rescue are covered by the term *technical rescue*?
7. In general, what is the minimum space requirement for a helicopter landing zone?
8. What action must be taken before approaching a helicopter with a rescue basket deployment?
9. What PPE should be worn during helicopter landing zone operations?
10. List three hazards that might be present during a post-hurricane response.

ACTIVITIES

1. Review your department's SOPs regarding the response to hazardous materials incidents and compare the SOPs to what really happens in the field. Do the responders comply? Are they trained to the level required for the response?
2. Access your department's response levels to technical rescue incidents. What services or types of responses are provided? What safety issues have been identified and what procedures are in place to reduce the hazards?
3. Access the potential for civil disorders and terrorism events in your community. What procedures are in place to ensure a safe response?
4. Review the types of natural disasters that are likely to occur in your response area. Do the procedures in place consider responder safety and health?

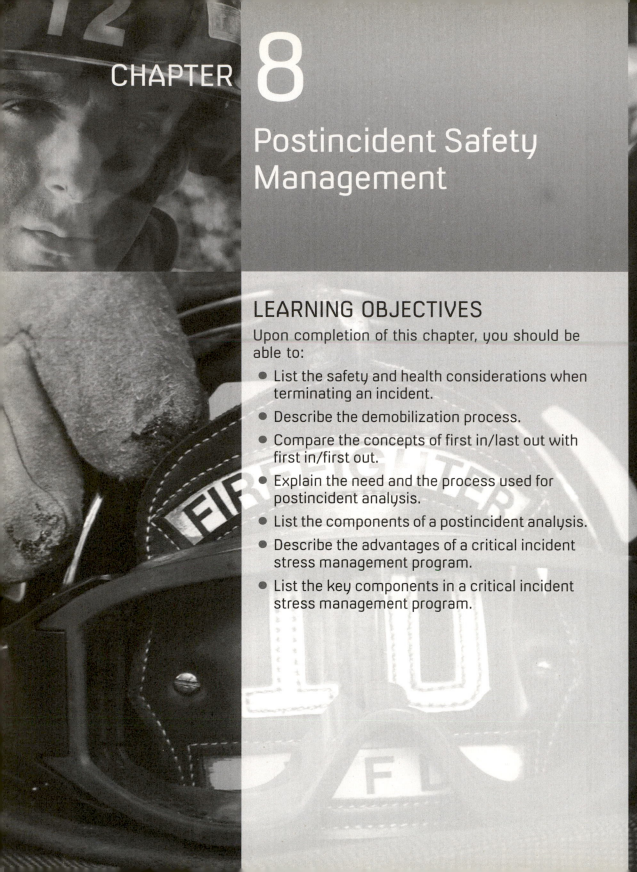

CHAPTER **8**

Postincident Safety Management

LEARNING OBJECTIVES

Upon completion of this chapter, you should be able to:

- List the safety and health considerations when terminating an incident.
- Describe the demobilization process.
- Compare the concepts of first in/last out with first in/first out.
- Explain the need and the process used for postincident analysis.
- List the components of a postincident analysis.
- Describe the advantages of a critical incident stress management program.
- List the key components in a critical incident stress management program.

On November 15, 2008, a fire fighter died after being crushed by a roof collapse in a vacant/abandoned building. Fire fighters had initially used a defensive fire attack to extinguish much of the fire showing from the second-floor windows on arrival. After the initial knockdown, fire crews entered the second floor to perform overhaul operations. During overhaul, the roof collapsed with several fire fighters still inside on the second floor. The victim and two other fire fighters were trapped under a section of the roof. Crews were able to rescue two fire fighters (who self-extricated), but could not immediately find the victim. After cutting through roofing materials, the victim was located by fire fighters, unconscious and unresponsive. He was removed from the structure and transported to a local hospital, where he was pronounced dead. Key contributing factors identified in this investigation include: dilapidated building conditions, incendiary fire originating in the unprotected structural roof members, inadequate risk-versus-gain analysis prior to committing to interior operations involving a vacant/abandoned structure, inadequate accountability system, lack of a safety officer, an inadequate maintenance program for self-contained breathing apparatus (SCBA), a poorly maintained and likely inoperable personal alert safety systems (PASS), and ineffective strategies for the prevention of and the remediation of vacant/abandoned structures and arson prevention.

NIOSH investigators concluded that to minimize the risk of similar occurrences, fire departments should:

- Ensure that the incident commander conducts a risk-versus-gain analysis prior to committing to interior operations in vacant/abandoned structures, and continues the assessment throughout the operations.
- Ensure SOPs are developed for fighting fires in vacant/abandoned buildings.
- Ensure that the incident commander maintains close accountability for all personnel operating on the fireground.
- Ensure that a separate incident safety officer, independent from the incident commander, is appointed at each structure fire.
- Ensure that a respiratory protection program is in place to provide for the selection, care, maintenance, and use of respiratory protection equipment, including PASS devices.
- Safety and health considerations continue both throughout the incident and after it is over.

Source: Information for this case study was taken in part from CDC/NIOSH Fire Fighter Fatality Investigation Report F2008-37. http://www.cdc.gov/niosh/fire/reports/face200837.html

INTRODUCTION

The last step in the safety and health program in terms of incident response is the consideration for postincident activities. Included in this chapter are suggestions for terminating the incident, postincident analysis, and critical incident stress management. Postincident safety and health considerations can be easily overlooked if not integrated into the total safety and health program. In other words, responders tend to let their guard down after the incident is over. Therefore, the safety officer and supervisors must be trained, and safety during this phase should be emphasized and committed to as part of the program.

The following Firefighter Life Safety Initiative is the focus for this chapter:

13. *Firefighters and their families must have access to counseling and psychological support.*

There are a number of postincident considerations that are decribed throughout this chapter.

INCIDENT TERMINATION

Incident termination can be comprised of three stages: demobilization, returning to the station, and postincident analysis. Postincident analysis is discussed in detail in the next section of this chapter.

NOTE Responders tend to let their guard down after the incident is over.

Demobilization

Demobilization is the stage in the incident when the incident commander in a large incident, with the recommendation of the planning section, evaluates the on-scene resources and compares them with the current situation. Using a fire scenario, demobilization usually occurs when the fire has been brought under control and some of the firefighter resources can be placed available and returned to stations. Demobilization, particularly at very large incidents, requires planning and coordination (see **Figure 8-1**). The National Interagency Incident Management System provides a form, Form ICS-221, to assist in the coordination of demobilization.

demobilization The process of returning personnel, equipment, and apparatus after an emergency has been terminated.

FIGURE 8-1
Demobilization is the
stage in which units and
personnel begin to leave
the scene.

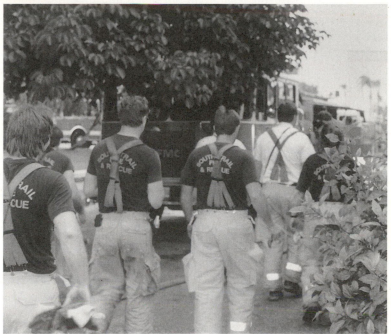

There are numerous safety and health issues that the incident commander should address during this stage that should be part of the department's operating procedures and safety policies.

One consideration is the sequence in which operating crews are released. The fire service has historically used the **first-in/last-out** approach. Very often, the unit that is first on the scene is the last unit to leave.

> **NOTE** Lately some departments have begun to change this approach, using first in/first out as a demobilization tool.

This process is good because it permits the units that arrived first, and probably have worked the longest and generally under the worst conditions, to be relieved first. Some arguments are made regarding this process as the first arriving apparatus are normally committed to firefighting and have ladders raised to the building or hose connected. This argument can be overcome when departments have **standardized apparatus**

first in/last out The common approach often used at an emergency scene; basically, the first arriving crews are generally the last to leave the scene.

standardized apparatus Apparatus that has exactly the same operation and layout of other similar apparatus in a department; for example, all of the department's pumpers would be laid out the same, operate the same, and have the same equipment, which is useful for the situations when crews must use another crew's apparatus.

and equipment. In this case it should make no difference that a first-arriving crew could later leave in another crew's apparatus. The first-in/first-out approach is a very important consideration during periods of extreme heat or cold and high-humidity or high-stress events.

There also is the concept of special call additional or fresh crews to perform the overhaul, or property conservation, function. The idea is that the fresh crews, not fatigued from fighting the fire, are more likely to wear appropriate PPE and be more mindful of hazards.

The incident commander and the planning officer should also consult with the on-scene safety officer during the demobilization process. The safety officer may be more aware of crews that need to be relieved or that have operated for extended periods. The safety officer can obtain some information from the rehabilitation officer who should have been monitoring the status of members of each crew.

The use of rapid intervention crews or teams (RIC) was discussed in Chapter 5. During demobilization, the incident commander should make an informed decision as to when the RIC is no longer needed. It is not uncommon for injuries to occur long after the initial fire fight is over. In fact, some disastrous collapses in which crews have been trapped and killed have occurred during the overhaul phase of a fire. One such incident is noted in the case review of this chapter.

Once released from the emergency scene or, in the case of a medical incident, once the call has ended, there are other safety concerns as well.

NOTE The supervisor or senior person assigned to the crew must ensure the readiness of the team for the next incident.

After any incident, the crew members should be questioned about injuries they may have incurred or any feelings about the incident that may prompt the supervisor to start the critical incident stress management process. In the case of an incident in which crew members have the potential to be exposed to infectious diseases, the crew should do as much personal decontamination as possible at the scene or at the hospital prior to returning to the vehicle. At this point the supervisor or company officer must make a decision as to whether the unit can be placed in an available status or whether the crew members or equipment are not ready. It is not uncommon in the emergency service field to check equipment and stock after a call to ensure readiness. The proper reloading and readying of equipment before leaving the scene also has safety considerations (see **Figure 8-2**).

The human safety elements cannot be overlooked in this process. The supervisor should also be alert to signs of **critical incident stress**.

critical incident stress Stress associated with critical incidents, such as the injury or death of a coworker or a child.

FIGURE 8-2
Safety is also a consideration during the trip back to the station and putting equipment back on the apparatus.

© Cengage Learning 2013

Returning to Station

Vehicle operators may be fatigued and tired from a prolonged incident and supervisors must ensure they are alert enough to drive apparatus back to the fire station. Reviewing the NFPA injury and death statistics, the category for vehicle crashes is titled "responding/returning." Injuries do occur during the returning-to-station stage of the incident. Returning to the station often provides the first chance for the crew to discuss the incident among themselves.

Once back at the station, the apparatus is made ready for the next emergency, supplies are replaced, and equipment cleaned. Again, the safety program must incorporate the human needs at this point. The supervisor should be alert for signs of fatigue and the crews should be assessed again for any injuries or illness that may be attributed to the incident. Remember, emergency service responders often do not complain, because they see minor injuries as part of the job. The supervisor should be alert to visual signs such as limping or favoring an arm or leg, cuts, bruises, and general differences in the responder. There are cases where line-of-duty deaths have occurred upon return to the station and the firefighter went to lie down. The supervisor should be aware and check on the responders after incidents to ensure their well-being.

Once the equipment and crews have been taken care of, the unit is readied for the next assignment. Before long, however, the supervisor should sit down and discuss the incident with the crew in an informal manner. This is the first step in the **postincident analysis (PIA)**.

postincident analysis (PIA) A critical review of the incident after it occurs; the postincident analysis should focus on improving operational effectiveness and safety.

POSTINCIDENT ANALYSIS

A postincident analysis should be done on any incident to which multiple units have responded and operated and on any incident involving a serious injury or death to civilians or responders. These incidents may include fires, technical rescues, hazmat incidents, multiple-casualty incidents, vehicle accidents, and victim entrapments. Often called a critical review, or critique, the postincident analysis (PIA) is a step-by-step look at what happened at the incident, including what went right and what went wrong. The components that the PIA should focus on are:

- Resources
- Procedures
- Equipment
- Improving operational effectiveness

The term *critique* is sometimes associated with "criticize," therefore the term *postincident analysis* is used in this text.

There are two steps of the PIA. The first is the informal discussion among the crew after the incident (see **Figure 8-3**). Although informal, this is a good time to review what each member saw and did and what the unit did as a whole. The supervisor during this discussion should make notes for use at the formal PIA. This process can—and should—also occur in volunteer organizations and may occur at the scene.

Usually within some fixed time after the incident, all crews that were present at the incident should get together for the formal PIA. This PIA must be attended by all those who carried out a specific task at or for the incident, including those members who filled positions within the incident management system (see **Figure 8-4**). Often the

---

FIGURE 8-3
The informal PIA may begin at the scene.

Courtesy of Palm Harbor Fire Rescue

FIGURE 8-4
The formal PIA should involve all members that operated at the incident.

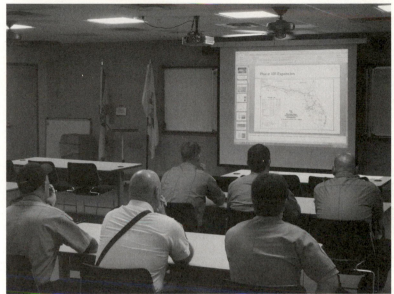

© Cengage Learning 2013

incident commander will call for and run the PIA. However, it is better in most cases to have an objective person other than the incident commander facilitate the PIA, because this then permits a free exchange of information and allows the incident commander to be more involved in the PIA as opposed to having to lead the analysis. The PIA events should be recorded for future changes in procedure or policies or to reaffirm current ones. Some departments videotape the PIA and then distribute the tape department-wide so that every member can learn from the incident. At minimum there should be a written form that is filled out and filed for each PIA conducted (see **Figure 8-5**).

A good procedure for the PIA is to have all the responders discuss—usually using a graphic or plot plan of the incident—what they did on arrival, what worked, and what did not work. If injuries or deaths occurred as a result of the incident, surrounding events should also be noted and investigated. Someone present at the PIA should have a copy of the SOPs available and compare the actions taken to those required by the procedures. For example, if the SOP says that the first-arriving ambulance will set up a treatment area and, in the case in question, the first-arriving ambulance entered the building to fight the fire, then either the SOP needs to be reevaluated or the crew should state the reasons for their actions.

The objective of the PIA should be to highlight positive and negative outcomes of an incident. The PIA can also provide a tool for future training and show a need for changes in procedures and policies or even a need for a new piece of equipment. The PIA should be a positive process, allowing all attendees to have a chance to speak and discuss concerns, particularly those related to short- and long-term safety and health issues. The PIA can also aid in the investigation process into any incident injuries and fatalities and become a documented reason to develop procedures that may prevent them in the future. See **Figure 8-6** for the relationship between SOPs, training, operations, and the PIA.

INCIDENT DATA

Date:

Alarm Time:

On scene time:

Response delays:

Location:

Units Responded:

Construction type:

Weather and Unusual Conditions:

Size up:

Water supply:

Location:

Distance:

Hose size:

Unit:

Pre-plan information available: Yes   No

Was it used?   Yes   No

COMMAND/CONTROL

Initial strategy:

Transfer of Command: Yes   No   From: To:

Initial Operational Mode:

Transition of Operational Mode (Off. to Def./Def to Off.) :

Use of Support Staff:

Effectiveness of Command Structure:

Interface with Other Agencies:

Staging:

Where:

Staging officer:

INCIDENT PRIORITIES

LIFE SAFETY

(Provide comment in all areas. If not applicable in this instance, mark N/A)

Courtesy of Palm Harbor Fire Rescue

FIGURE 8-5

A Sample Postincident Analysis – Structure Fire form.

FIREFIGHTER SAFETY

Use of accountability:

Readiness level of R.I.T:

Safety Officer effectiveness:

Care and rehab of our people:

Security of utilities:

Identification of hazards:

Risk assessment:

SEARCH/RESCUE

By whom, what was found:

Primary Search:

Secondary Search:

Evacuation:

Occupant services:

INCIDENT STABILIZATION

Exposure Protection:

Line Placement:

Company:

GPM/Length/Nozzle:

Proper GPM:

Confinement:

Forcible entry:

Where:

How:

Line Placement:

Company:

GPM/Length/Nozzle:

Proper GPM:

Extinguishment:

Company:

FIGURE 8-5 Continued

Courtesy of Palm Harbor Fire Rescue

GPM/Length/Nozzle:

Proper GPM:

Ventilation:

Location:

Method:

Overhaul:

PROPERTY CONSERVATION

Salvage:

LESSONS LEARNED

1. The three things that went very well would include:

   A.

   B.

   C.

2. The three things that we, as a department, would like to have done better on would include:

   A.

   B.

   C.

3. The greatest safety lesson that came from this included:

4. If we had to identify one area of training that we could all stand some refresher training on, it would be:

Other Comments:

Personnel Present:

Courtesy of Palm Harbor Fire Rescue

FIGURE 8-5 Continued

FIGURE 8-6

The relationship between the PIA, SOPs, training, and operations.

© Cengage Learning 2013

CRITICAL INCIDENT STRESS MANAGEMENT

Stress is a normal part of everyday life; however, too high a stress level leads to negative reactions such as actual physical illness, job burnout, and lack of productivity. In terms of safety and health, a stress management program is necessary, including provisions for employee assistance programs as discussed in Chapter 4. This section of the text focuses on post–critical incident stress management and the associated health and safety considerations.

Emergency responders are expected to tolerate certain levels of stress. However, some events that are usually high in stress levels include those that are powerful in terms of emotion, or a combination of many smaller events. These events are called *critical incidents*. Many years ago, responders were expected to handle these events on their own and were often criticized if they displayed weakness or emotion toward these critical events. Fortunately, this is no longer the case, and not only has critical incident stress management become accepted in the emergency service arena, but also it is a part of most organizations today.

Events that typically result in critical incident stress include the following:

- Traumatic death or severe injury to a coworker, particularly those suffered in the line of duty (see **Figure 8-7**)
- Suicide of a coworker
- Traumatic death or serious injury to children
- Mass casualty events
- Prolonged events
- Death or injury to a person caused by the emergency responder; for example, a traffic accident while responding to an incident
- Events in which there is a great deal of media attention
- Personally significant events; for example, the death of an elderly man when the responder just lost an elderly family member

When a critical incident occurs, it is important that a process be in place both for defusing stress and for debriefing. This process should include access to the Critical Incident Stress Management (CISM) system. In some cases, this debriefing should take place before the responders have to ask for it, as the responders may not feel they need it or may not ask for fear of showing signs of weakness. Each organization sets procedures for when and how to access the system. Some systems automatically set the process in motion when any of the critical incidents listed takes place or at the request of an incident commander or company supervisor for any event.

---

stress The body's reaction to an event; not all stress is bad; in fact, some level of stress is necessary to get a person to perform; for example, the stress associated with a report that is due is often the motivating factor in doing it.

FIGURE 8-7
Accidents involving coworkers are particularly stressful. CISM should always follow these incidents.

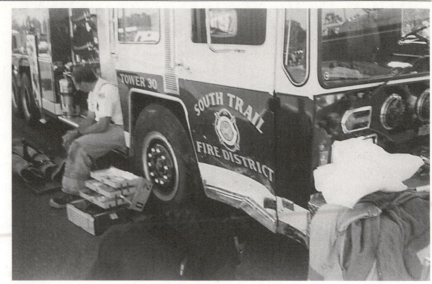

© Cengage Learning 2013

Without CISM, the responder may be more likely to develop negative physical, behavioral, and psychological reactions. Unwanted physical reactions include fatigue, sleeplessness, changes in eating habits, and body aches. Behavioral changes may also occur, including changes in activity levels, difficulty in concentration, nightmares, flashbacks, memory problems, and isolation. Psychological symptoms include fear, guilt, sensitivity, depression, and anger. In some extreme cases, if not properly managed, a person suffering from critical incident stress reactions may begin to abuse alcohol and drugs and may resign from employment in the emergency service field. The safety and health program that includes CISM helps the responder deal with the issues associated with the event and hopefully avoid many of the unwanted reactions.

There are different types of CISM, both at the scene and after the event. The first should occur at the scene and is termed **peer defusing**. Peer defusing involves an informal discussion about the event by a group of peers who experience the event together. This phase often occurs normally as the incident de-escalates and responders get together and discuss the event. The second is the formal debriefing that usually occurs sometime after the incident, maybe even a few days later, and this is the formal process in which only the responders to the incident get together with a peer debriefer and discuss their roles at the incident and the overall response.

peer defusing The concept of using a trained person from the same discipline to talk to an emergency responder after a critical incident serves as a means to allow the responder to talk about his or her feelings about the event in a nonthreatening environment.

NOTE Two issues are very important in this phase:

1. All those involved should attend.
2. The peer counselor must be educated in the process and in this type of operation.

Very often CISM teams are regionalized, which allows for a peer debriefer from the same occupation but an outside agency to serve in that capacity. It would be difficult for a department to have an inside team, as each debriefing would involve a peer from the same department who was not at the incident. This approach is not good. Good CISM team characteristics include regionalization, a mental health professional as a team member, being nonpartisan, and having received appropriate training in CISM.

NOTE Total confidentiality and relative anonymity are vital elements of a debriefing, and the peer debriefer must follow and buy into these concepts, as should the management of the response organizations.

A third type of CISM is on-site defusing, which can be useful at large, long-duration events. As they leave the operational area, responders are evaluated by a member of the local CISM team for signs and symptoms of distress and are treated appropriately. The fourth type of CISM is demobilization. Demobilization allows responders, after operating at a large-scale event, to have a buffer period of 30 to 40 minutes before leaving the location. This period can provide a time for nourishment and a brief education on the signs and symptoms of critical incident stress.

Although the incidents have occurred forever, it was not until the 1980s that incident stress was recognized as a problem. The safety and health of the department members and their longevity in the response system depend on CISM. It is a necessary component to any safety and health program.

SUMMARY

Safety and health after the incident are sometimes overlooked. During this phase of the response, there is potential for injuries to occur and, therefore, this must be considered in the safety and health SOPs. Postincident safety considerations begin with the termination of the incident, at which time the incident commander makes a decision about the demobilization of resources, and these resources return to their stations. On a single unit medical-type incident, the incident is terminated once the patient has been handed over to the emergency department or to a transport provider. It is important during this phase of the operation that the same consideration be given to the human components of the crew as is given to the equipment and supplies.

Postincident analysis (PIA) and critical incident stress management (CISM) are part of the postincident safety considerations. The PIA can begin when the unit returns to the station and the superiors have an informal discussion with the crew regarding the incident. Later, a formal PIA must be conducted using a prescribed format and documented for future use in training or for updating SOPS. All personnel who responded to the particular incident must be included in the formal PIA.

Another postincident safety and health concern is critical incident stress management. Critical incident stress management programs are integral to the safety and health program. Teams should be available to assist members following incidents that generate high levels of stress. Often these teams are formed on a regional basis to allow for peer reviewers from outside of one's own agency. Supervisors must have training in and be aware of common incidents that can lead to critical incident stress reactions. All personnel must be aware of available CISM programs and how to access them at any time, given the need.

Concluding Thought: Many essential safety and health program functions occur after the incident itself, but they are just as important as those that occur before or during an incident.

REVIEW QUESTIONS

1. What are the three stages in terminating the incident?
2. Differentiate between the first-in/first-out and the first-in/last-out approaches. List advantages and disadvantages of each.
3. It is best to allow the incident commander to facilitate a PIA.
   A. True
   B. False
4. Using a supervisor from within the organization as a peer debriefer is an important step in the CISM process.
   A. True
   B. False
5. List three qualities of a good CISM team.
6. The RIC can be released once a fire is declared under control.
   A. True
   B. False
7. Describe the relationship between training, SOPs, operations, and the PIA.
8. As much as possible, decontamination should occur before leaving the scene.
   A. True
   B. False
9. Who should be present at a formal PIA?
10. The informal PIA should occur on return to the station before the unit is placed back in service.
    A. True
    B. False

ACTIVITIES

1. Review your department's SOP for demobilization at emergency incidents. Which process is used, first in/first out or first in/last out? Consider possible changes to your department's approach or SOPs. Is the approach used in your department working? What roadblocks might prevent a first-in/first-out procedure?

2. Develop a postincident analysis SOP for your department. If there already is one, how effective is it? Are the objectives noted in this text being achieved?

3. Research information on CISM teams in your area. Have personnel in your department had training in recognizing critical incident stress reactions and in how to access the CISM team?

CHAPTER 9

Personnel Roles and Responsibilities

LEARNING OBJECTIVES

Upon completion of this chapter, you should be able to:

- List the roles and responsibilities of the individual responders and their relationship to the overall safety and health program.

- List the roles and responsibilities of the supervisors and their relationship to the overall safety and health program.

- List the roles and responsibilities of the emergency service management team and their relationship to the overall safety and health program.

- List the roles and responsibilities of the incident commander and his or her relationship to the overall safety and health program.

- List the roles and responsibilities of the safety program manager and his or her relationship to the overall safety and health program.

- List the roles and responsibilities of incident safety officers and their relationship to the overall safety and health program.

- List the roles and responsibilities of the safety committee and its relationship to the overall safety and health program.

On May 22, 2010, a fire fighter died while conducting a search in a residential house fire after vomiting, removing his facepiece, and inhaling products of combustion. A captain and the victim had entered the 6,000 square foot residential structure with an uncharged 1¾" hoseline to perform search and rescue operations for an elderly occupant and a dog, while an attack crew began fire suppression operations in another part of the structure. After locating and extricating the dog, the captain and the victim continued the search in increasingly heavy black smoke. After he vomited, clogging his nose cup, the victim became separated from the captain. The victim tried to clear his mask and verbally called out that he was in trouble. The captain called a Mayday and immediately began searching for him. Two rapid intervention teams (RITs) also searched for the victim. The victim was found approximately 11 minutes later and approximately 24 feet from where he was last seen. The RIT crew removed him from the structure to the front yard, where paramedics performed medical care. The victim was transported to the local medical center where he was pronounced dead. After the incident it was determined that the elderly occupant was not at home. Factors that contributed to this death were:

- Fire fighter became ill, causing a self-contained breathing apparatus emergency and a separation from his captain.
- The location of the victim was not immediately known.
- Fire growth contributed heavy smoke, zero visibility, and heat conditions.

Recommendations:

- Develop, implement, and train on a procedure that addresses what to do if the self-contained breathing apparatus becomes inoperable due to a clogged nose cup, such as with vomitus.
- Ensure that fire fighters are trained on primary search and rescue procedures that include maintaining crew integrity, entering structures with charged hoselines, and following hoselines in low visibility.
- Ensure that fire fighters are trained and retrained on Mayday competencies.
- Ensure that staffing levels are appropriate to perform critical tasks.

Every member of the emergency service organization has a specific role to play in the safety and health program, from individual responsibilities to those of the entire organization.

Source: Information for this case study was taken in part from CDC/ NIOSH Fire Fighter Fatality Investigation Report F2010-13. http://www .cdc.gov/niosh/fire/reports/face201013.html

INTRODUCTION

As part of putting a safety program together, the safety program manager must understand the various roles that the members of the organization play. Who exactly is responsible for the safety and health program within an organization? After reviewing the information that has been presented thus far, the answer should be clear—everyone!

NOTE Safety and health responsibility extends from the new individual member to the top of the organization, whether that top is the chief or an other-titled director.

Recalling the SAFEOPS approach from Chapter 1 (see **Table 9-1**), it can quickly be seen that many different persons in the organization have responsibility for one or more components of SAFEOPS. The interplay of the various groups and individuals in the safety and health program can be compared to a continuous circle, and that circle is only as strong as its weakest member (see **Figure 9-1**).

The roles of every member in the organization can be easily linked to the following Firefighter Life Safety Initiatives:

1. *Define and advocate the need for a cultural change within the fire service relating to safety; incorporating leadership, management, supervision, accountability, and personal responsibility.*
2. *Enhance the personal and organizational accountability for health and safety throughout the fire service.*
3. *Focus greater attention on the integration of risk management with incident management at all levels, including strategic, tactical, and planning responsibilities.*
4. *All firefighters must be empowered to stop unsafe practices.*

Although there are differing responsibilities in the various organizational levels, the goal of the program remains the same. This chapter examines these various levels and the associated roles that the persons at these levels may play.

TABLE 9-1 Primary Responsibility for the Components of the SAFEOPS Approach

COMPONENT	PRIMARY RESPONSIBILITY
Supervision	Incident commanders
	Supervisors
	Safety program manager
	Management
	Incident safety officers
Attitude	Individual responders
	Supervisors
	Management
	Incident commanders
	Safety program manager
	Incident safety officers
	Safety committee
Fitness	Individual responders
	Supervisors
	Management
	Incident commanders
	Safety program manager
	Incident safety officers
	Safety committee
Education/Wellness	Individual responders
	Supervisors
	Safety program manager
	Management
Organizational involvement	Safety committee
	Safety program manager
	Management
Procedures	Individual responders
	Incident safety officers
	Incident commanders
	Safety committee
	Safety program manager
	Management

(continues)

TABLE 9-1 Continued

COMPONENT	PRIMARY RESPONSIBILITY
Standards/Regulations (compliance with)	Incident safety officers
	Incident commanders
	Safety committee
	Safety program manager
	Management

© Cengage Learning 2013

FIGURE 9-1

The interrelationship of roles can be likened to a continous circle, with the strength of the circle dependent on the strength of each component.

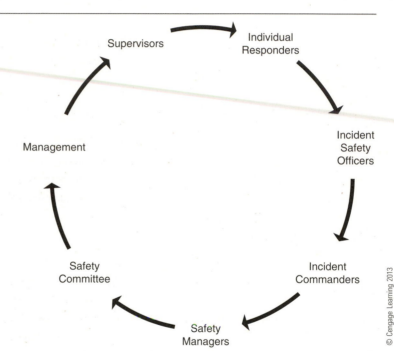

© Cengage Learning 2013

INDIVIDUAL RESPONDERS

Individual responders are probably the most important link in the safety and health program. This may sound unusual; however, the individual responders can have in fact a great impact on the entire program.

NOTE Regardless of how many policies are written, how many safety officers are assigned, or how much money is put into a program, without compliance and a positive safety attitude on the part of the responders, the program will not be effective.

The "A" in SAFEOPS represents attitude. Attitude may well be the most important obstacle to overcome when dealing with safety and the individual responder. Although changing, the historical belief that injuries and death simply go with the occupation is still part of many responders' attitudes. The belief or attitude that "it will not happen to me" is also human nature. There was once a line in a war movie in which the actor was shot. The actor's line was "I'm shot. I never thought I would get shot!" One would think that a person fighting a battle in a war could assume that the potential for being shot was reasonably high, yet he was surprised that it really happened to him. We see this same belief with many people, whether it involves a crime, car accidents, or, in some cases, being injured at work. In order for a responder to want to operate safely and comply with procedures and policies, the responder must realize that the potential for injury to him or her is real and could actually happen.

Individual responders must also realize that they are members of a team. Depending on the type of response unit, this may be a two-person team on an ambulance or a three- or four-person team on a fire unit. Although an individual, the responder's actions almost always have some impact on the other members of the team. For example, if a firefighter does not wear all of the necessary protective equipment and is overcome with smoke, the rest of the team has to rescue the firefighter, and the entire team is affected.

The individual responder should also be willing to take an active role in the safety and health program. This may involve making suggestions, working on the safety committee, and using good safety practices. It is important that the more senior responders set an example, particularly when training a new recruit. **Text Box 9-1** describes some individual responder do's and don'ts.

SUPERVISORS

Although many members of the organization fall into the category of supervision, for the purposes of this chapter, the term "supervisor" is used to describe first-level or frontline supervision in the emergency response organization. For fire service departments this may be a company officer, lieutenant, or captain assigned to a firefighting unit. In the case of sole emergency medical services (EMS) providers, it may be a senior paramedic or crew chief. Regardless of title, the safety and health roles and responsibilities for the first-line supervisor remain the same (see **Figure 9-2**).

NOTE The first-line supervisor is the person who supervises the team.

The supervisor ensures that team members stay together, maintains accountability, sees that seat belts are fastened prior to apparatus moving, and ensures that personal protective clothing for the job at hand is worn. The supervisor may also be responsible for inspection of protective clothing and seeing that regular maintenance procedures are followed.

TEXT BOX 9-1

DO'S AND DON'TS FOR INDIVIDUAL RESPONDERS

Do be:

- *An active team player.* Be prepared to fill your role, including following safety-related procedures, and watch out for other team members. If a team member does not have protective gloves on when starting an IV, remind him or her.

- *A good communicator.* This applies to both listening and speaking. When working as a team, communicate with other team members. For example, when raising a ladder or lifting a stretcher, maintain contact with other team members and ensure that the team makes the lift together on command.

- *A leader.* Display safety leadership. Set the example.

- *Aware of surroundings.* Maintain a constant awareness of surroundings for both yourself and the team. Look for danger signals, communicate concerns.

- *A responder who works within individual abilities.* If you know you cannot do something alone, get help. Countless back injuries associated with lifting could have been avoided when others were available to help.

- *Together.* Stay with your assigned team; no freelancing.

- *A responder with a Safety Attitude!*

Don't be:

- *Preoccupied.* Emergency responders are called on at a moment's notice. Be thinking about the incident and the potential outcome. Put everything else out of your mind.

- *Complacent.* Do not let the routine nature of a medical incident allow you to let your guard down. Do not assume that the fourth fire alarm today in the same building is not a fire.

- *Surprised.* Many events that occur at emergency incidents are predictable—a disorderly crowd in a bar, a building collapse after exposure to fire. Predict and be prepared.

- *A responder who attempts to do things outside of personal abilities.* If you cannot swim, do not jump in the water to rescue someone.

- *Maintain an attitude that it will not (cannot) happen to me.* Maintain a Safety Attitude.

© Cengage Learning 2013

FIGURE 9-2
Fire department company officers have a number of responsibilities associated with the safety of their crews.

NOTE The first-line supervisor is the link between the organization's management and the individual responder.

The supervisor is responsible for seeing that procedures are followed and are understood by team members. The supervisor also communicates safety concerns from the team to higher management levels in the department.

The first-line supervisor generally has risen through the ranks and completed some type of performance-based promotional process. In some cases, supervisors might be assigned based solely on seniority. In either case, the first-line supervisor should have the necessary mix of experience and education to recognize dangers inherent in the occupation, whether on an emergency scene, while responding, or in the station. The supervisor must apply this knowledge and take responsibility for the crew in terms of safe operations, including enforcing organization standard operating procedures, following laws, and maintaining awareness of team location and any dangers present. The supervisor must also follow the practices noted previously for individuals, with specific emphasis on leadership and setting the example.

EMERGENCY SERVICE MANAGEMENT

Somewhere in the organization above the first-line supervisory level is the organization's management. This might be the fire or EMS chief, or the director of emergency services, depending on the type and structure of the organization. For the purpose of this section, emergency service management is used to define the very top levels of the organization, or senior management staff. The safety program manager, regardless of rank or position, should be a part of the senior management structure (this is discussed later in this chapter in the Safety Program Manager section).

Senior management also has many roles in the safety and health program. While having roles and responsibilities both as individuals and as supervisors, the senior management staff in addition often controls the financial resources and gives final approval for policy and procedure implementation.

The senior management staff must give the safety program priority in terms of support, both financial and administrative. This support must be realized by the individual responders at lower organizational levels. For example, the paramedic on the street must know that the EMS chief has a commitment to provide a safe work environment with the employees' health as a driving factor. Have you ever heard the statement, "We will do such-and-such because it is required by OSHA"? A better statement might be, "We will do such-and-such, because, after analysis, we have found it to be another measure of safety provided to our members." One way of communicating this commitment is through a written organizational safety policy. An example policy is shown in Text Box 9-2.

INCIDENT COMMANDER

Approximately 45% of firefighter injuries occur at the fireground. The standards, and in some cases the regulations, require that incidents, regardless of type, be managed with an IMS. At the highest level of the IMS is the incident commander (IC). The fact that most injuries occur at the incident scene and that the IC is in charge of the incident scene leads to the conclusion that the IC has a great deal of responsibility in ensuring safety at the incident. The IC is charged with setting the overall incident strategy and strategic goals and assigning tactics to meet these goals (see Text Box 9-3). The IC must balance risks involved in meeting the strategic goals while maintaining a constant regard for team safety. A careful—but quick—risk benefit analysis must be performed before deciding on a strategy.

The IC must first decide on the basic strategy—offensive or defensive—to be employed. The IC should make this determination based on past experience, education, and risk assessment. The IC must rely on information that may or may not be accurate at the time he receives it. Common information includes the type of building, potential for occupants, the progression of the fire, and the resources available. For example, a vacant building with moderate fire involvement might immediately lead one to believe that a defensive attack is the strategy of choice. But if this building were in the middle of a block of closely built, occupied buildings, an offensive attack might be ordered.

Incident commanders must have a good knowledge of fire dynamics, building construction, and accepted tactics. They must also have a good level of field experience. Incident commanders can rely on past incidents to provide knowledge as to what is effective and what is ineffective. They can also learn from attending postincident analyses

TEXT BOX 9-2

ORGANIZATIONAL SAFETY POLICY

Management Commitment and Involvement Safety Policy

It is the policy of our department to follow the highest safety and health standards. Safety does not merely occur by chance, but occurs as the result of commitment and participation. Our firefighters are our most important assets. Lost time from the job due to accidental injuries is costly to everyone. Our objective is a safety and health program that will eliminate accidents and injuries or reduce their number to an absolute minimum.

The responsibility for a safe and healthful environment is shared throughout the fire department. The safety officer provides my ongoing direction and guidance in safety matters. I, in conjunction with the department officers, am responsible for training. All officers/supervisors shall execute and enforce these safety rules, policies, and procedures with the utmost regard for the safety and health of our firefighters.

Our department's policy is that employees make suggestions and report accidents and unsafe conditions to their supervisor and that they know their rights not to perform unsafe work tasks, without fear of reprisal. I provide the leadership for the safety and health program's effectiveness and improvement to promote safe working conditions through employee feedback, safety committee reviews, and program funding. The safety officer responsible for the safety and health program for our organization is:

Ongoing safety program activities shall include:

- A safety committee with firefighter representation
- Safety orientation, education, and training
- Safety committee review of safety rules, accidents, policies, and procedures
- Appropriate corrective actions on accident reports and revising safety rules when appropriate

Firefighters are responsible for following all aspects of the safety and health program and for compliance with all rules and regulations. Responsibility for continuously practicing safety while performing their duties is a condition of employment. All violations of this policy shall be investigated and disciplinary action applied as warranted.

Fire Chief: _____   Date: _____

TEXT BOX 9-3

STRATEGIC DIFFERENCES

The incident commander must evaluate the risks of offensive and defense fire attacks.

- *Offensive Attack Strategy.* Firefighters are in close contact with the fire and exposed to all the inherent dangers, including burns, falling objects, being lost, being overcome with toxic gases, and building collapse.

- *Defensive Attack Strategy.* Firefighters are outside the immediate area of fire danger. Most protective equipment is still required. Dangers still exist, such as building collapse, falling objects, and exposure to heat and fire products.

© Cengage Learning 2013

of other incidents and applying the lessons learned to future incidents. Although the decision-making process taught in many officer programs relies on a several-step decision-making process, in actuality, research shows that ICs use what is known as the recognition-primed decision-making (RPD) process at emergency scenes. Basically the RPD theory says that ICs do not have time at an emergency to go through a several-step process with selection of alternatives; instead, they generally arrive and immediately begin to take control of the incident based on past experience with similar incidents. If something worked ten other times at a bedroom fire, it will work again. For the IC to utilize this process and to provide for sound strategic plans, simulation training is necessary, so the IC can be trained to react to predicable events at the emergency scene.

SAFETY PROGRAM MANAGER

The safety program manager is the focal point for safety-related activities in the organization. As stated previously, the safety program manager should be a member of the senior staff. The safety program manager has the overall responsibility for overseeing the program and all associated components. *NFPA 1521* requires that, even though the fire chief has the responsibility for the department's safety and health program, a health and safety officer shall be assigned to manage the fire department's occupational safety and health program.

The roles and responsibilities of safety program managers are summarized in **Text Box 9**-4. Depending on the size of the organization, the safety program manager may have a staff of assistants, or, in a smaller organization, may fulfill these roles in addition to other duties, such as being the training officer. In either case, the safety program manager must have sufficient experience and education in risk management principles, cost/benefit evaluations, the safety and health issues that responders face, infection

TEXT BOX 9-4

THE ROLE AND RESPONSIBILITIES OF THE SAFETY PROGRAM MANAGER

- Be the organization's risk manager.
- Be a member of senior staff with direct access to the chief or director.
- Receive and act on recommendations from individuals and the safety committee.
- Cause safety and health policies to be developed.
- Be liaison with workers' compensation providers.
- Be liaison with the department physician on health-related issues.
- Investigate injuries or line-of-duty deaths.
- Investigate all accidents and near misses.
- Maintain all records relating to health and safety.
- Participate in and develop safety and health training programs.
- Perform facility inspection for unsafe conditions.
- Evaluate procedures from a safety and health perspective.
- Perform cost/benefit analyses.
- Attend and participate in postincident analyses.
- Have input into departmental standard operating procedures.
- Have input into the specifications of apparatus and equipment to ensure compliance with applicable safety and health standards.
- Perform evaluation and analysis of data relating to safety and health.
- Maintain awareness of trends in emergency service safety and health, including new standards or regulations and court decisions.

© Cengage Learning 2013

control, and emergency service operations to be fully effective. In addition, *NFPA 1521* requires that the safety program manager meet the requirement of Fire Officer Level I in *NFPA 1021*.

INCIDENT SAFETY OFFICER

Because most of the injuries that emergency service workers experience occur at the incident scene, the assignment and function of the incident safety officer is key to the safety and health program.

NOTE Depending on the size and type of the organization, the incident safety officer may be a dedicated position from within the safety division, or may be a first-line supervisor assigned to the safety role on an incident-by-incident basis.

Some departments assign extra units to large incidents and have these crews function in certain safety-related roles, including incident safety officer, accountability, entry control officer, or rapid intervention companies. In some cases, the incident safety officer is in name only as, for example, when the assignment is made to anyone who is free at the incident. This is not good practice, because the incident safety officer must have additional knowledge, be well experienced with the incident at hand, and have no other duty. Different incidents may require that a different level of safety officer be assigned (see **Figure 9-3**). For example, a competent incident safety officer at a fire incident may not be a very good safety officer at a hazardous materials emergency. The incident and the level of expertise of the individuals should dictate the assignment.

NFPA 1521, Standard for Fire Department Safety Officer, states the requirements and responsibilities of the incident safety officer and requires that one be appointed when the need occurs. The requirements in *NFPA 1521* for the incident safety officer include that that person shall meet Fire Officer I per *NFPA 1021*, have the experience and education to manage incident scene safety for the type of incident, have a knowledge of safety and health hazards involved in emergency operations, know building construction, know the department's personnel accountability system, and understand incident scene rehabilitation.

After arrival at the incident scene, the incident safety officer should evaluate the incident and what is happening. A prediction of what could or is going to happen should also be part of this evaluation. This evaluation involves a 360-degree walk around the scene and examination of available information, such as preincident plans. The incident safety officer should talk to occupants or owners and question them about hazardous situations relating to the property or vehicle—for example, alternative fuel

FIGURE 9-3
An incident safety officer
at a vehicle accident.

use in a car, or gunpowder in a home, or the storage of explosives. In Florida a firefighter was killed by a .22 caliber rifle subjected to the intense heat in a house that was on fire.

Depending on the type of incident, the incident safety officer should assess the operation from a safety point of view and relay findings to the incident commander. Although all members operating at the scene have the responsibility to report and stop unsafe acts, the safety officer must be given the specific authority to instantly stop unsafe acts that are immediately dangerous to responders. The stopping of an assignment must be relayed to the IC promptly as well. During the incident evaluation, the incident safety officer should assess the operating personnel, including the use of proper protective clothing, accountability, and crew intactness. **Text Box 9-5** examines some incident types and the role of the incident safety officer. One good technique to remind the safety officer of what to look for at different type of incidents is to utilize a clipboard and worksheet that can be completed. These should be durable enough to be used under various weather conditions. There should be a worksheet for each different type of incident that the department may be called on to respond to. **Figure 9-4** is an example of a worksheet for multiunit response to a vehicle accident with entrapment. This worksheet can later be used as part of the postincident analysis.

The incident safety officer serves as a member of the IC's command staff and must be a resource for them. The IC must use the safety officer's knowledge and expertise in helping with strategic and tactical decisions.

THE SAFETY COMMITTEE

One of the most common methods to involve employees and to have a forum for employees to share concerns about the organization's safety and health program is through the use of a safety committee. The safety committee can be a resource for the safety program manager and other safety professionals in the organization. Some safety committees, as a group, review accidents and injuries and assist the program manager with efforts to reduce them. Having employee involvement on the committee helps to deliver to all employees the message that safety is a priority and gives the employees the feeling that they are involved, through representation, in formulating procedures and safety-related policies (see **Figure 9-5**).

If the members of an organization are represented by a labor union, often there is a contractual requirement to have a safety committee and to have employee representation on it.

NOTE Even if there is no contractual requirement, the safety committee should have representation from all levels of the organization.

Do not forget to have all administrative personnel represented, including people in nonresponse roles, such as dispatchers or office staff.

TEXT BOX 9-5

VARIOUS INCIDENT TYPES AND THE ROLE OF THE INCIDENT SAFETY OFFICER

Fire Incidents

- Matches strategy to the situation. Ensures the strategy is being followed by all operating members.
- Makes sure crews are intact.
- Makes sure IMS is in place.
- Makes sure the accountability system is in place.
- Makes sure proper level of protective equipment is being used.
- Determines status of building or vehicle structure.
- Establishes collapse zones and other hazard zones.
- Directs rehabilitation setup.
- Checks physical condition of personnel.
- Continually monitors radio communication.
- Ensures proper scene lighting.
- Communicates unsafe locations such as holes in floors, wires down, and backyard swimming pools to operating teams.
- Makes sure there are means of egress for crews.
- Determines whether risk assessment has been done.
- Determines that rapid intervention crews are ready.
- Makes sure utilities are secured.
- Starts investigation of any on-scene accident or injury.
- Participates in postincident analysis (PIA).

EMS Incidents

- Makes sure IMS is in place if multiple units are operating.
- Makes sure accountability system is in place if needed.
- Determines that protective equipment is available for the task at hand.
- Ascertains structural condition of crashed vehicles, e.g., broken glass, sharp metal.
- Ascertains traffic hazards and creates a buffer zone.
- Reduces exposure to bloodborne pathogens.
- Ensures safe number of rescuers, i.e., not doing too much with too little.

(continues)

- Stops leaking fluid.
- Calls for fire suppression support.
- Initiates infection control.
- Ensures time for rehabilitation.
- Checks physical condition of crews.
- Starts investigation of any on-scene accident or injury.
- Participates in PIA.

Hazardous Materials Situations

- Makes sure IMS is in place.
- Makes sure accountability system is in place.
- Makes sure rapid intervention team is available and with the same protective clothing as entry team.
- Ensures teams are operating within level of training and available resources.
- Makes certain teams have adequate protective clothing.
- Establishes proper zones.
- Monitors pre- and post-entry physical conditions.
- Ensures that rehabilitation is available.
- Makes sure decontamination process is adequate.
- Starts investigation of any on-scene accident or injury.
- Participates in PIA.

Technical Rescues/Special Operations

- Makes sure IMS is in place.
- Ensures accountability system is in place.
- Ensures there are adequate resources, both human and equipment.
- Makes sure personnel are acting within level of training.
- Ensures protective equipment is available.
- Ensures proper equipment is available—ropes, air supply, personal flotation devices, and so forth.
- Starts investigation of any on-scene accident or injury.
- Participates in PIA.

FIGURE 9-4
The incident
safety officer's
checklist for a
vehicle accident.

Fire/EMS Department
Incident Safety Officer Worksheet
Vehicle Accidents

Scene
- ☐ Traffic Controlled
- ☐ Utilities Secured
- ☐ Hazardous Materials

Vehicles
- ☐ Stabilized
- ☐ Structural Stability
- ☐ Cargo
- ☐ Fuel Type

Operations
- ☐ IMS
- ☐ Fire Supression Support
- ☐ Extrication
- ☐
- ☐
- ☐
- ☐

Personnel
- ☐ Adequate
- ☐ Proper Training
- ☐ PPE
- ☐ Teams Intact
- ☐ Within IMS
- ☐ Accountability
- ☐ Rehabilitation
- ☐

Site Sketch

Units Assigned
_____
_____
_____
_____
_____
_____

Safety Issues
_____
_____
_____
_____
_____
_____
_____

© Cengage Learning 2013

Once formed, the safety committee should elect a chairperson who may or may not be the safety program manager. The committee should meet as often as required to do business, but no less than four times a year. The minutes of the meetings should be published in a format that allows all members of the department to read and review them. Any recommendations from the committee should be forwarded to top management for action. Depending on the organization, the safety committee chairperson may have direct access to the top manager without going through the various levels of command. **Text Box 9-6** shows an example of safety committee policy.

FIGURE 9-5
The safety committee must have members from all levels of the organization.

© Cengage Learning 2013

TEXT BOX 9-6

EXAMPLE OF A SAFETY COMMITTEE POLICY

Safety Committee

This section formally establishes the safety committee, provides a clear statement of duties and responsibilities, and outlines administrative, procedural, and reporting requirements. The safety committee will meet at minimum quarterly.

Organization and Duties

The duties and responsibilities of this committee are to:

1. Actively participate in safety and health training programs and evaluate program effectiveness.

2. Review the results of inspections to detect unsafe conditions, materials, practices, and environmental factors.

3. Review and analyze all workplace accidents for hazards, trends, and proper corrective actions and recommend updates to safety rules and procedures as needed.

4. Recommend methods and activities aimed at hazard reduction and/or elimination.

5. Review, compile, and distribute hazard, safety, and health information to the department.

6. Review suggestion or incentive programs.

7. Review new laws, regulations, and guidelines to assess impact and develop courses of action to effectively meet and comply with requirements.

(continues)

Membership

Participation on the safety committee will include, as a minimum, the following assigned members from our organization:

The safety committee chairperson is _____.
(May be one of the positions below.)

Position	Name of Assigned Member
Supervisor	_____
	_____
Firefighter	_____
	_____
Safety officer	_____
	_____
Non-Responder Representative	_____
	_____

Order of Business

The order of business for the safety committee meeting will be:

1. Record of attendance
2. Approval of previous minutes
3. Unfinished business/open action items
4. Review of accidents, equipment failures, and recalls
5. Presentation of new business
6. Reports on special assignments
7. Reports of inspections
8. Suggestions, awards, recognitions, and incentives
9. Special feature (film, talk, demonstration)

Record Keeping

Formal documentation shall be kept for each meeting. Confidentiality is critical in the discussion of incidents and injuries and must be addressed both in the discussion and in the writing of the minutes. Copies of safety committee minutes are to be posted in all work locations and are to be forwarded to the fire chief as well as safety committee members.

FIGURE 9-6
Like teamwork at an emergency, each member of the organization has an integral role in the safety and health program.

© Cengage Learning 2013

SUMMARY

Each member of the organization has a particular responsibility in the safety and health program. From the individual responder's role in maintaining teamwork and following procedures to top-level management appropriating money for safety equipment, the program will not be effective without everyone doing his or her part and assuming his or her role (see **Figure 9-6**).

Some specific roles that must be considered in the program are the safety program manager, the incident commander, and the incident safety officer. The safety program manager is the nucleus of all activities relating to the program and must be knowledgeable in multiple subjects dealing with safety and risk management. The incident commander also plays a vital role as the person who selects the strategy and tactics for a given situation. A commitment to safety and an understanding of risk management are important attributes for the incident commander. The incident safety officer responds to incidents and serves as a member of the incident commander's staff. The incident safety officer performs an evaluation of the incident and the operations and recommends operational changes to the incident commander. The incident safety officer must have the knowledge and experience specific to the particular incident at hand. There are various safety-related issues that are particular to the incident type.

The safety committee is an extension of the organization's membership and safety manager. The safety committee provides a forum in which members make suggestions on safety and health matters. The committee can also be used to investigate accidents and injuries and make recommendations for improving existing or implementing new procedures designed to reduce occupational injuries. The safety committee should have membership and representation from all levels of the organization, and may or may not have direct access to the organization's senior manager.

Concluding Thought: Every person in the organization plays an integral part in the overall safety and health program. Everyone has a specific role with a defined responsibility.

REVIEW QUESTIONS

1. According to this text the first-line supervisor would be responsible for:
   a. Ensuring that proper protective clothing is worn at a single-unit medical incident.
   b. Communicating safety concerns on the crew's behalf to the next level of management.
   c. Ensuring that safety procedures are followed while responding to an incident.
   d. All of the above are responsibilities of the first-line supervisor.
2. List three safety and health–related attributes of an effective individual responder.
3. List three don'ts or things that an individual responder should avoid.
4. Attacking a room and contents fire from the inside of a structure would be considered an offensive fire attack.
   a. True
   b. False
5. A _____ fire attack would be one in which the responders protect exposed structures and do not enter the fire building because of fire conditions.
6. List three things that an incident safety officer might look for at a fire incident.
7. List three things that an incident safety officer might look for at an EMS incident.
8. List three things that an incident safety officer might look for at a hazmat incident.
9. List five roles of the safety program manager.
10. Describe the relationship between the incident commander and the incident safety officer.

ACTIVITIES

1. Compare the activities of your department's safety committee to the recommendations in the text. What could be improved? If you do not have a safety committee, how could one be started?

2. What are the procedures in your department for assignment of an incident safety officer? What improvements can be made? How well does the incident safety officer interface with the incident commander?

3. Examine your own roles at incidents. Do you find yourself displaying any of the "don't" traits listed in the chapter? What can you yourself do to improve the safety of your team?

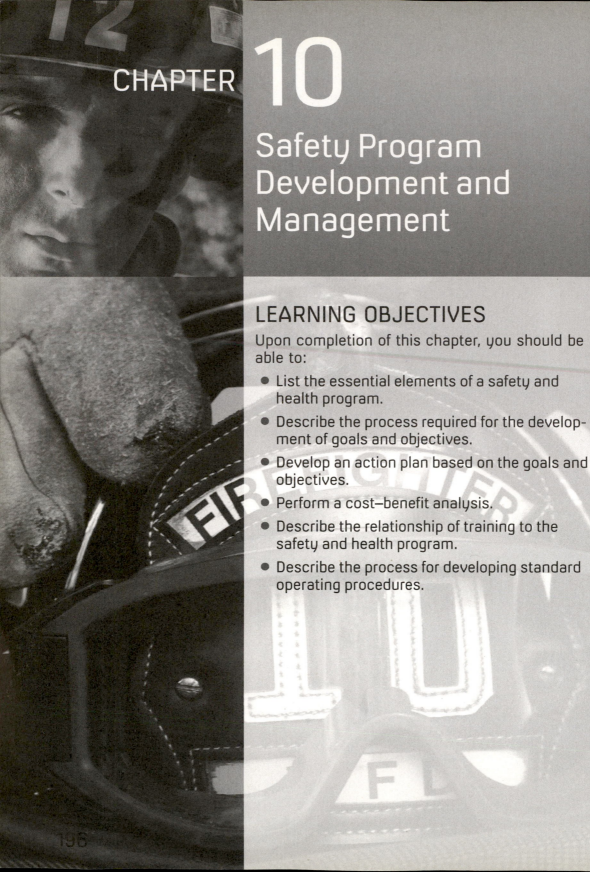

CHAPTER **10**

Safety Program Development and Management

LEARNING OBJECTIVES

Upon completion of this chapter, you should be able to:

- List the essential elements of a safety and health program.
- Describe the process required for the development of goals and objectives.
- Develop an action plan based on the goals and objectives.
- Perform a cost–benefit analysis.
- Describe the relationship of training to the safety and health program.
- Describe the process for developing standard operating procedures.

In May 2009, two captains, a lieutenant, and five fire fighters were injured during a natural gas explosion at a strip mall in Maryland. At 1254 hours, dispatch reported a natural gas leak inside a business at a strip mall. Five minutes later, the initial responding crew and the incident commander (IC) arrived on scene to find a gas company employee looking for an underground gas leak. Approximately 6 minutes later, a natural gas leak was found near the exterior rear corner of the structure. After 23 minutes on scene, approximately 45 civilians were evacuated from 7 occupied businesses. A captain exited the rear door of the business that had called in the natural gas leak and noticed fire along the roof line. Crews in the front and rear of the structure had begun to pull hoselines as another captain was looking out the rear doorway of a middle unoccupied business and noticed the electric meter located on the exterior wall was on fire. Anticipating an explosion, he tried to leap out the rear doorway. At the same time, a fire fighter had entered the front door of the unoccupied business, noticed the heavy smell of natural gas, and felt air rush by as the structure exploded. Debris and fire blew out the front, rear, and roof of the structure. The captain who tried to leap out the rear doorway was blown into the rear parking lot and the fire fighter who had entered the front of the structure was blown out the front door and covered with debris. Numerous other fire fighters, primarily near the front of the structure, were blown off their feet and hit with debris. An uninjured captain issued a Mayday, followed by the IC ordering evacuation tones and a personnel accountability report. Crews began to look for the captain who was blown out the rear doorway. He had walked around the side to the front of the structure, and radioed his location to command. Fire fighters began moving injured personnel to ambulances staged in the front parking lot. Eight fire fighters and a gas company employee were transported to local hospitals. The injuries ranged from third degree burns to an ankle sprain.

Key contributing factors identified in this investigation included: insufficient execution of the fire department's updated standard operating guidelines (SOGs) on incidents involving flammable gas, e.g., apparatus and fire fighters operating in a flammable area (hot zone); the accumulation of natural gas in the structure's void spaces; unmitigated ignition source; insufficient combustible gas monitoring equipment usage and training; and ineffective ventilation techniques. NIOSH investigators concluded that, to minimize the risk of similar occurrences, fire departments should:

■ Ensure that standard operating guidelines for natural gas leaks are understood and followed.

■ Contact utility companies (natural gas and electric) immediately to cut external supply/power to structures when gas leaks are suspected.

■ Ensure that gas monitoring equipment is adequately maintained and that fire fighters are routinely trained on proper use.

■ Ensure that ventilation techniques are conducted after ignition sources are mitigated.

■ Ensure that rapid intervention teams are staged at the onset of an incident.

■ Ensure that collapse/explosion control zones are established when dealing with a potential explosion hazard.

In this case a total of eight members were injured in one incident. The direct cost of replacing these members on duty and the indirect cost of the lost time must be considered in program cost/benefit analysis.

Source: Information for this case study was taken in part from CDC/ NIOSH Fire Fighter Fatality Investigation Report F2009-13. http://www .cdc.gov/niosh/fire/reports/face200913.html

INTRODUCTION

This text has thus far allowed the reader to form a good understanding of what is required for a comprehensive safety and health program. Having presented the information as tools, the text now must lead the reader down the path to getting started. With all this information, it is difficult to determine just where to start.

This chapter provides just that information—what the essential elements are to a successful program, the setting of goals and objectives, action planning, overcoming barriers, performing a cost–benefit analysis, relating the training program to the goals of the safety program, and developing standard operating procedures. Although presented in this order for the purpose of the chapter, the actual order of the process may be different, depending on your organization.

The information presented in the chapter is related to four Firefighter Life Safety Initiatives:

1. *Define and advocate the need for a cultural change within the fire service relating to safety; incorporating leadership, management, supervision, accountability and personal responsibility.*

3. *Focus greater attention on the integration of risk management with incident management at all levels, including strategic, tactical, and planning responsibilities.*

9. *Thoroughly investigate all firefighter fatalities, injuries, and near misses.*

11. *National standards for emergency response policies and procedures should be developed and championed.*

ESSENTIAL ELEMENTS

The first step in the process of developing the safety and health program is to determine what essential elements are needed. This may vary by department type, size, and nature of responses. In some cases, one role may be shared by more than one person; in others, one person may perform more than one role. Clearly, using the information presented in this text and applicable standards such as *NFPA 1500*, the following elements are essential to an effective program:

- Top management that is committed to the program
- A safety and health program manager
- A comprehensive risk management plan
- A safety and health policy
- Some type of record-keeping system for data analysis
- Incident safety officer(s)
- A training program
- Standard operating procedures
- Proper equipment and PPE that meets requirements and standards
- A safety committee
- A department physician
- An accident and injury investigation process
- Infection control program
- Provisions for critical incident stress management
- Access to local, state, and national injury and death statistics
- A process to analyze and implement policies to comply with local, state, and national regulations and standards

Safety is not cheap, but it is often cheaper than one serious injury. Some of the elements, such as a department physician, can be expensive, whereas getting the commitment of top management is free. However, the program can be developed over a period of time as funding becomes available. It is better to get something started and then seek out the funding for the more expensive components. At a minimum, we need the commitment of top management, a safety program manager, a safety committee, a record-keeping system, and access to data in order to get the program off the ground.

SETTING GOALS AND OBJECTIVES OF THE PROGRAM

Once the essential elements of the safety program are in place and the risk identification process is completed, the safety program manager and safety committee start setting goals and objectives for the program. This goal setting is accomplished through a common approach taught in many management courses and may be considered the organization's strategic planning for safety.

Goals

The first step in this process is to determine **goals**. Goals are defined as broad-based statements with a measurable outcome and time frame. Goals can be developed by the safety program manager working alone or with the safety committee. However, using the group process and the safety committee will probably result in better acceptance by the members, and the group process generally provides a better result. There may be any number of goals set forth, and these become the road map to guide the safety program.

NOTE It should be cautioned that the number of goals developed for a particular time frame should be limited, and that each goal should be realistic and obtainable.

If there is too much to do in the specified time frame or if goals are never obtained, the organization's members may quickly lose interest and the program may suffer.

For the purpose of applying a realistic example in this chapter, we will assume that the organization's safety committee has identified a particular problem with back injuries. After the problem was identified, the committee began goal development. After discussion, consensus was reached and the following goal recommended:

Safety Program Goal 1: To develop a comprehensive back injury reduction program to reduce the number of job-related back injuries by 50% within the first 12 months of implementation, with an additional 25% reduction each 12 months thereafter.

In analyzing this goal, it is determined that the goal is broad enough, does not give any specifics, is measurable (50% reduction), and has a specified time frame (12 months). This would be a good goal statement for any back injury reduction program.

Once the goal statement has been developed, more specific statements of action must be developed in order to meet the goal. These specific statements, or objectives, provide the road map to reach a goal.

NOTE Whereas a goal is broad based, an objective is specific, but also must be measurable and within a given time frame.

goals Broad statements of what needs to be accomplished.

Objectives

There is no magic number of objectives needed to reach a goal. Instead, it depends on the complexity of the goal and the number of interrelated activities needed.

NOTE An objective must only deal with one specific activity and, as with goals, be obtainable within the given time frame.

The following is a poorly written objective based on our fictitious back injury problem:

Objective 1.1: A training program will be developed and presented to all employees, and each employee will be issued a back support belt within 30 days.

An analysis of this objective reveals a couple of flaws. First, the objective deals with more than one interrelated issue, i.e., training and support belts; and second, the time frame set forth may not be reasonable. The four following objectives are written to comply with the suggested guideline and are designed to meet the back injury reduction program goal:

Objective 1.1: Within 90 days, develop a standard operating procedure requiring assistance when performing nonemergency lifting of equipment over 50 pounds and train employees in the application of the procedure.

Objective 1.2: Within 120 days, form a team to research options for relocating heavy equipment on the apparatus to lower levels and submit a written report with recommendations to fire department management.

Objective 1.3: Within 120 days, consult with the department physician to develop a fitness routine designed to strengthen back muscles and incorporate education on proper lifting into daily fitness training.

Objective 1.4: Within 180 days, research the use of back support belts in emergency service applications, including organizational results, and file a written report to the safety committee including recommendations and a cost–benefit analysis.

These objectives are all clear and to the point. They are measurable and include reasonable time frames. Each could be easily analyzed at the end of the specified time frames for completion. See **Text Box 10-1** for a summary of the goal and objectives for the example program.

The goal and objective process is dynamic and changing from two perspectives: First, the goals and objectives have to be developed for each problem area and may change over time as new problems are identified. Second, as part of the process, each goal and objective should be reevaluated during the implementation process to see whether the objective is being met and, if not, what has to be changed.

TEXT BOX 10-1

SUMMARY OF GOAL AND OBJECTIVES FOR THE BACK INJURY EXAMPLE

Problem: The average annual frequency of back injuries in the fire department is higher than that in comparable departments.

Safety Program Goal 1: To develop a comprehensive back injury reduction program to reduce the number of job-related back injuries by 50% within the first 12 months of implementation, with an additional 25% reduction each 12 months thereafter.

- *Objective 1.1:* Within 90 days, develop a standard operating procedure requiring assistance when performing nonemergency lifting of equipment over 50 pounds and train employees in the application of the procedure.

- *Objective 1.2:* Within 120 days, form a team to research options for relocating heavy equipment on the apparatus to lower levels and submit a written report with recommendations to fire department management.

- *Objective 1.3:* Within 120 days, consult with the department physician to develop a fitness routine designed to strengthen back muscles and incorporate education on proper lifting into daily fitness training.

- *Objective 1.4:* Within 180 days, research the use of back support belts in emergency service applications, including organizational results, and file a written report to the safety committee including recommendations and a cost–benefit analysis.

© Cengage Learning 2013

NOTE Safety program goals and objectives should be published and recognized by all members of the organization.

This can be accomplished through the safety committee meeting minutes or any other department-wide distribution process.

ACTION PLANNING

As described in the previous section, the goal statements are the road map for the safety program and the objectives are the road map to the goal. What, then, is the road map to the objectives? The answer is the action plan (see **Figure 10-1**). The action plan is a step-by-step written guide to meeting an objective. Each objective should have an action plan developed.

FIGURE 10-1
The relationship between goals, objectives, and the
action plan.

RISK IDENTIFICATION

GOALS

OBJECTIVES

ACTION PLAN

© Cengage Learning 2013

The action plan has several components to be considered. The action plan should list the goal and objective that it is designed for and list the team members' names. The action plan should be developed by the team that has been assigned the objective and should contain very specific step-by-step actions, often set up in tabular form. For each step, the action, a completion time benchmark, the person responsible, the resources needed, any support or roadblocks anticipated, and a completion date should be examined and included. The only column that is sometimes left blank is the support/roadblock column, although generally support and roadblocks should be identified during the action planning process. **Table 10-1** is an example action plan for Objective 1.2 of our back injury reduction plan.

As seen in Table 10-1, the action plan covers all of the necessary steps to fully implement the objective. Further progress can easily be measured throughout by using the completed benchmarks. Certain support and roadblocks have been identified so the team can anticipate these issues during the implementation process. This action plan example, although maybe not all inclusive, is a good example of a road map to the objective.

COST–BENEFIT ANALYSIS

Although the action plan presented in Table 10-1 did not include a cost–benefit analysis, very often one is necessary to support a position, particularly if the objective involves any financial outlay. A cost–benefit analysis is often used to show that the initial outlay for a program will save in future reduction of risk. Examples might include the purchase of new firefighting PPE and the reduction in burn injury costs, or the cost of an extensive driver training program and the associated reduction in vehicle accidents.

Often when using a cost–benefit analysis, some assumptions must be made and sometimes the data are based on estimations of improvement rather than actual results. Unfortunately it is almost impossible to determine the exact outcome of a

TABLE 10-1 Sample Action Plan

Safety Committee
ACTION PLAN

Date: January 1, 2012

Team Members: Jones, Smith, Big, Little

Goal 1: To develop a comprehensive back injury reduction program to reduce the number of job-related back injuries by 50% within the first 12 months of implementation, with an additional 25% reduction each 12 months thereafter.

Objective 1.2: Within 120 days, form a team to research and implement options for relocating heavy equipment on the apparatus to lower levels and submit a written report with recommendations to fire department management.

Step No.	Action	Date to Be Completed By	Person Responsible	Resources Needed	Support(S)/ Roadblocks(R)	Date Completed
1	Review SOP adoption guidelines	1/10/12	Jones	None	None	
2	Contact other similar departments for SOPs	1/15/12	Smith	Telephone/ E-mail/Fax	(S) Chief association/ (S) Statewide labor organization	
3	Develop SOP	1/30/12	Smith	Computer	(S) Safety committee	
4	Determine heavy equipment	1/10/12	Big/Little	None	None	
5	Determine compartmentation options	1/30/12	Big/Little	Vehicle compartment inventory	(S) Company officers	
6	Have SOP draft approved	2/15/12	Smith	None	(S) FD Staff (R) FD Staff	
7	Begin training on SOP	2/20/12	Smith/Jones	Overhead training time	(R) Personnel acceptance	
8	Move equipment as recommended	3/1/12	Big/Little	None	(R) Personnel acceptance	
9	Full implementation	3/30/12	Team	None	None	
10	Evaluate outcome Recommend improvements	3/30/13	Team	None	(S) FD Staff	

program prior to its implementation. However, by using information from similar agencies that have done similar projects, an estimation of expected outcomes can be formed. Another way to view outcome expectation is to analyze historical injury data and determine which injuries would likely have been prevented if the program in question had been in effect. Another issue with cost–benefit analysis is that you cannot put a price on some losses that the emergency services incur. For example, the loss of one life is too much, and therefore a program would have to be undertaken regardless of the resultant cost–benefit analysis result.

A cost–benefit analysis typically allows the safety program manager to evaluate the cost-effectiveness of a program. The analysis examines the current cost of the risk and compares those costs to the cost of program implementation, considering both direct and indirect losses and costs. Recall from Chapter 3 that the cost of a risk is measurable and can be both direct and indirect. Direct cost might include the costs of medical treatment, the overtime paid to cover a vacancy on a crew, or the cost of replacing equipment. Indirect costs include loss of productivity, the loss of using the equipment, stress-related concerns of coworkers, and possibly the cost of replacing the employee. Direct and indirect costs may also be applied as program costs.

NOTE The first step in the cost–benefit analysis is to describe, numerically, the cost of the risk currently.

Again, using our back injury example, a study can be performed and can describe what is occurring at the present time, without intervention. The next step is to determine the cost of the risk after the intervention has been implemented. After this step, the manager can determine whether this measure would be effective from a purely cost standpoint. Finally, the cost of the program implementation has to be calculated. Once these three areas have been calculated, an informed decision can be made. Remember that a program that may only save $1,000 per year and cost $5,000 to implement may still be a good program because of the organizational benefit after the fifth year. **Figure 10-2** is an example spreadsheet using the back injury reduction example. The analysis is for illustration only and is not to be construed as showing exact costs.

An examination of Figure 10-2 reveals that the organization would save $24,360 the first year. Aside from some ongoing training and replacement back supports, the cost of the program in future years would be greatly decreased. Therefore, the savings over time would be higher and could be calculated using the same format.

TRAINING

There is a direct and close relationship between training and safety (see **Figure 10-3**). In fact, in many organizations the training officer or division has assumed the safety program functions. Many of the training programs offered have a basis in operating safely. Many of the safety initiatives that a department undertakes are disseminated through

COST/BENEFIT ANALYSIS

GOAL: Development of Back Injury Prevention Program—First Year

CURRENT SITUATION AND COSTS

25 Back injuries per year with resultant hospital/doctor visit average cost
$2,000 per visit and an average 1.5 days (36 hours) of work lost

Direct Costs

Medical expenses	$50,000

Indirect Costs

Overtime to cover vacancies on shift, at average hourly rate of $10.00/hr.

$36 \times 25 \times 1.5 \times \$20.00 = \$27,000$	$27,000
Total Costs	**$77,000**

FUTURE ESTIMATION OF SITUATION

13 Back injuries per year with resultant hospital/doctor visit average cost
$2,000 per visit and an average 1.5 days (36 hours) of work lost

Direct Costs

Medical expenses	$26,000

Indirect Costs

Overtime to cover vacancies on shift, average hourly rate

$36 \times 13 \times 1.5 \times \$20.00 = \$14,400$	$ 14,400
Total Costs	**$40,040**
First Year Savings (Estimation Based On Research)	**$36,960**

PROGRAM COSTS

Direct

Physician's time	$4,000
Back supports	$5,000

Indirect

Training time	$2,000
Apparatus compartmentation change	$ 1,600
Total Program Costs	**$12,600**

TOTAL COST/BENEFIT—FIRST YEAR

First Year Savings minus Program Implementation Costs	$24,360

FIGURE 10-2

An example cost–benefit analysis of a back injury prevention program.

FIGURE 10-3
Live Fire Training for
flammble liquid fires.

© Cengage Learning 2013

the department's training programs. From the back injury program example, a key component involved the training of members.

As presented in Chapter 2, a number of standards and regulations dictate how an emergency service organization operates. Within these standards and regulations are requirements for certain levels of training. A very good example is the hazardous materials incident operations regulation, which bases training requirements on five levels of response.

NOTE Training mandates are designed in hopes of ensuring that a minimum level of training is provided to responders in order that the responders can handle the incident safely and recognize dangers.

Many states require a minimum level of training for firefighters or paramedics. Within these training requirements, one can always find some requirement related to safety, whether it be use of SCBA or infection control.

Training is also important as the organization's safety program develops. The training program can be a vehicle for introducing and testing new procedures. Remember, the best accountability system on paper that does not work during training evolutions will probably not work at a real incident. Training may also be developed to help deal with existing injury problems, such as lifting injuries or thermal burns.

More than any other function within an emergency service organization, the training staff should be the most integrated into the safety program, both in development and in implementation.

TEXT BOX 10-2

NFPA STANDARDS RELATING TO FIRE DEPARTMENT TRAINING

NFPA 1401, Recommended Practice for Fire Service Training Reports and Records

NFPA 1402, Guide to Building Fire Service Training Centers

NFPA 1403, Standard on Live Fire Training Evolutions

NFPA 1404, Standard for Fire Service Respiratory Protection Training

NFPA 1410, Standard on Training for Initial Emergency Scene Operations

NFPA 1451, Standard for a Fire Service Vehicle Operations Training Program

© Cengage Learning 2013

Vast resources are available for the organization's training program including fire and EMS professional organizations, the United States Fire Administration, state fire/EMS training centers, local colleges, universities, the fire academy, textbooks, and the NFPA (see **Text Box 10-2**). Using the Internet for research (described in more detail in Chapter 12) can provide training information, ideas, and experiences from other emergency service departments.

DEVELOPING STANDARD OPERATING PROCEDURES AND SAFETY POLICIES

Developing safety procedures and policies is necessary in order to meet some of the goals and objectives defined in the program. After development, approval, and implementation, they must be reviewed for effectiveness and updated as necessary. An excellent resource for the development of SOPs is a USFA publication titled "A Guide to Developing Effective Standard Operating Procedures for Fire and EMS Departments." This guide basically describes a development process based on the following four steps:

1. Conduct a needs assessment
2. Develop the SOP
3. Implement the SOP
4. Evaluate the SOP

In starting the process, the committee developing the SOP or policy should review the applicable goal or objective that the policy has to satisfy, or, in other words, conduct a needs assessment. Other organizations may be contacted for copies of their SOPs for the same objective. A gap analysis can also be performed, which simply answers the question of where we are versus where we should be.

Once the needs assessment has been completed, the committee can develop the SOP based on organizational format. It is important that once a draft is developed, it is sent out for comment. It is a good idea to get comments from other members of the department and from various ranks within the organization. If the SOP or policy is specific to one work function or one workplace, it is very important to target the affected groups for comments. This inclusion allows for greater input and helps with member buy-in, once adopted. Getting feedback may also tell the committee that the procedure may not work in real applications.

Once this entire process has produced an approved new or changed SOP or policy, all members of the organization should be trained in its application. This is the implementation step. This step—which requires realistic training, not just a classroom reading—will tell if the procedure is practical for operations. Further feedback should be provided to ensure the procedure will work.

After the implementation step, there must be evaluation of the SOP, which should include feedback that can be used to update or adjust the SOP as needed. **Figure 10-4** graphically shows this process as described in the USFA publication.

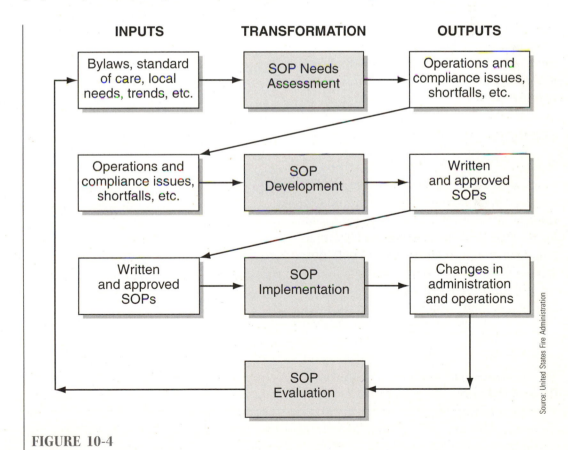

FIGURE 10-4
USFA SOP process.

SUMMARY

In order to be effective, the safety and health program must have several essential elements including a top management that is committed to the program; a safety and health program manager; some type of record-keeping system for data analysis; an incident safety officer or officers; a training program; standard operating procedures; proper equipment and PPE that meets requirements and standards; a safety committee; a department physician; and access to local, state, and national injury and death statistics. Some of these items may come after program development. However, at a minimum the program should have the commitment of top management, a safety program manager, a safety committee, a record-keeping system, and access to data.

Once risks are identified and analyzed, the safety program manager, in conjunction with the safety committee, must develop goals and objectives designed to minimize the risks. Goals are broad-based statements that are measurable and realistic and provide a guide to the safety program. Objectives are more specific and are the guides to meeting the goal. There may be one or many objectives for each goal. An action plan is developed based on each objective. The action plan is a step-by-step guide to meeting the objective. Each of these steps—the action, a completion time benchmark, the person responsible, the resources needed, any support or roadblocks anticipated, and a completed-by date—should be examined and included in the action plan.

For most goals and objectives, a cost–benefit analysis should be performed. The cost–benefit analysis examines the current situation in terms of direct and indirect cost. The costs of the risk after program implementation and the expected cost of the program are compared to the current cost, so that an informed decision can be made regarding implementation.

Training and safety are very closely related and somewhat dependent on each other. In some organizations the training division actually performs the safety function as well. Many training programs are designed for educating the members on how to perform tasks safely. Development of safety- and health-related SOPs and policies are also integral to the program. These can be developed by the safety committee after a gap analysis, a review of the safety goal and objectives, and contact with other organizations with similar policies. Once a draft is developed, it is important to get feedback from all levels of the department and to train all members on the implementation.

Concluding Thought: The safety and health program has several components. Using these components, the safety program manager goes through various processes to implement the program.

REVIEW QUESTIONS

1. A department physician is an essential component to have in place before a safety program can be established.
   a. True
   b. False
2. A goal is a broad statement with measurable results.
   a. True
   b. False
3. An action plan can be considered the road map to a goal.
   a. True
   b. False
4. List the four-step process for developing an SOP as outlined in the USFA publication "A Guide to Developing Effective Standard Operating Procedures for Fire and EMS Departments."
5. Which of the following is applicable to a cost–benefit analysis?
   a. The cost of the safety gear to be purchased.
   b. The current direct cost of the injuries if the program were not introduced.
   c. The indirect costs associated with the program after implementation.
   d. All of the above information would be necessary.

ACTIVITIES

1. Select an injury problem within your department. Write one goal and at least two objectives to develop a program to reduce or eliminate the problem.
2. Write an action plan for each objective.
3. Perform a cost–benefit analysis on the program.
4. Review your department's safety program and its relationship to the training division. Are they working together? Do they have similar goals? Is it effective?
5. Review your department's SOP development process. How does it compare to that presented in the text?

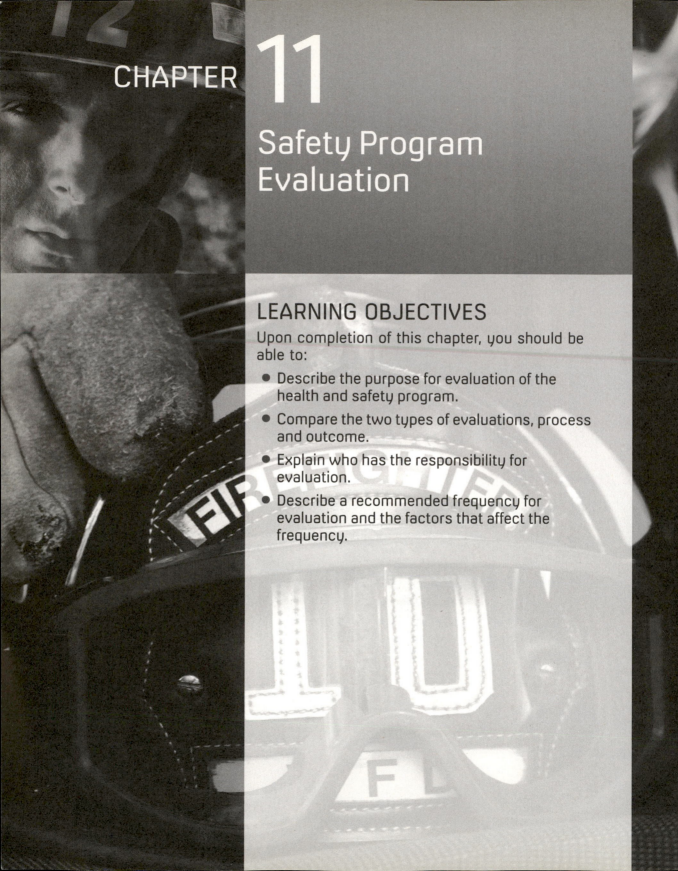

CHAPTER

11

Safety Program Evaluation

LEARNING OBJECTIVES

Upon completion of this chapter, you should be able to:

- Describe the purpose for evaluation of the health and safety program.

- Compare the two types of evaluations, process and outcome.

- Explain who has the responsibility for evaluation.

- Describe a recommended frequency for evaluation and the factors that affect the frequency.

In December 2009, a career fire fighter/paramedic was ejected from the front passenger seat of a fire department ambulance during a rollover incident. The ambulance was responding with lights and siren to a car fire with injuries. A motorist in a privately owned vehicle (POV) turned into the path of the ambulance as it was passing the POV. The ambulance struck the POV, lost control, and left the right side of the roadway. The ambulance overturned three times, and the victim was ejected. The driver sustained minor injuries. The victim's use of a seat belt could not be determined with certainty. The victim was airlifted by helicopter to a local trauma center where blunt force chest injuries and underlying medical conditions contributed to his death two days later. Factors that contributed to this incident included:

- Failure of the motorist to yield the right-of-way to an approaching emergency vehicle with audible and visual signals in use.
- Ejection of the victim.
- Blunt chest trauma resulting in multiple rib fractures.
- The victim's underlying medical conditions.

Recommendations:

- States and authorities having jurisdiction should take steps to ensure that motorists are aware of, understand, and follow state traffic codes/laws pertaining to yielding the right-of-way to approaching, authorized emergency vehicles using audible and visual signals.
- Fire departments should take steps to ensure that seat belts are properly worn at all times.

Even the best of safety programs will not be effective unless they are continuously evaluated and a change in behavior has been achieved.

Source: Information for this case study was taken in part from CDC/ NIOSH Fire Fighter Fatality Investigation Report F2009-30. http://www .cdc.gov/niosh/fire/reports/face200930.html

INTRODUCTION

No health and safety program is complete without some sort of evaluation process. In fact, *NFPA 1500*, Chapter 4, requires that the risk management plan be evaluated. Once the program has been designed and implemented, a process must be undertaken to evaluate its effectiveness. This evaluation may take any number of forms, depending on the goal or objectives being evaluated. Furthermore, the time frame for evaluation will also be goal dependent.

NOTE Simply put, the method of evaluation should compare where the organization is now compared to where it was prior to program implementation.

In order to be effective, an understanding of the evaluation process and the measuring of results must be undertaken. Once this knowledge is gained, the process can be included in the program (see **Figure 11-1**).

Program evaluation is linked in part to the following Firefighter Life Safety Initiative:

9. *Thoroughly investigate all firefighter fatalities, injuries, and near misses.*

THE EVALUATION PROCESS

The process of evaluation may take different forms; however, it usually involves the comparison of statistics from one time period with another time period. The results can then be compared to what was expected. Differences between expected outcomes and actual outcomes should be evaluated for program changes. There are three reasons to evaluate the program:

1. To see if the program is effective.
2. To determine the response to the program from the members' perspective.
3. To facilitate program changes.

FIGURE 11-1

A safety program manager analyzes data for the annual evaluation.

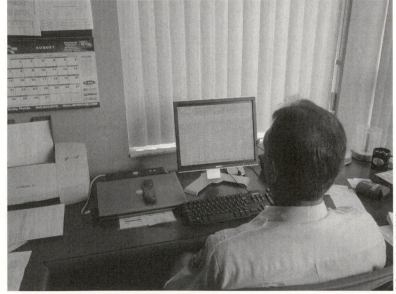

© Cengage Learning 2013

There are two methods of evaluation that should be used: **process evaluation** and **outcome evaluation**.

process evaluation	The ongoing evaluation of the various processes associated with a program or task.
outcome evaluation	An evaluation that answers the question, "Did the program meet the expected goals?"

Process Evaluation

Process evaluation, **Figure 11-2**, is an analysis of the procedures of the program and is undertaken throughout the program. Basically, the process evaluation answers the question, "How well did the processes in the program do what they were intended to do?" For example, the back injury prevention program designed in Chapter 10 had a component involving training on proper lifting techniques. A process evaluation for this component would examine whether the training had the desired effect. Simply put, are the responders using the proper lifting techniques they were trained to do?

The following questions are necessary in the process evaluation:

■ Who was affected by the program?
■ To what extent were they affected?

FIGURE 11-2
The process evaluation continuum.

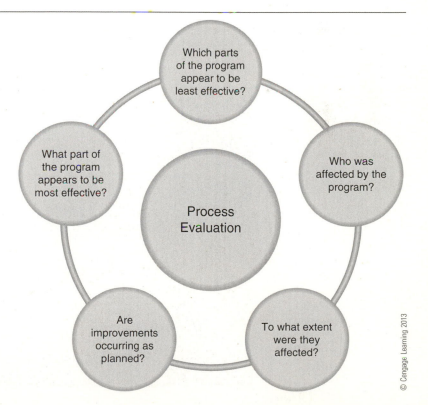

Which parts of the program appear to be least effective?

What part of the program appears to be most effective?

Who was affected by the program?

Process Evaluation

Are improvements occurring as planned?

To what extent were they affected?

© Cengage Learning 2013

- Are improvements occurring as planned?
- Which parts of the program appear to be most effective?
- Which parts of the program appear to be least effective?

There are a number of ways to perform the process evaluation including written or practical testing on safety policies and procedures, visual observation at incident scenes and around the station, and feedback from supervisors. Process evaluation can be used to measure results against program objectives. Process evaluation will tell the safety program manager what program changes are needed to increase the effectiveness of the program.

Outcome Evaluation

Outcome evaluation, **Figure 11-3**, is measured after the program has been in effect for a while. This evaluation examines what the program did. The outcome evaluation answers the question of whether the program is being effective. Again using the back injury prevention program as an example, an outcome evaluation would be used at the end of a time period, probably a year, to determine what effect the program had. Did the program actually reduce back injuries?

FIGURE 11-3
The outcome evaluation continuum.

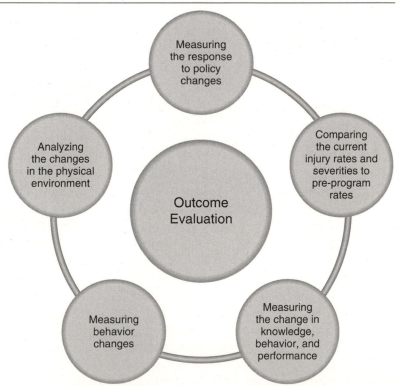

© Cengage Learning 2013

Several areas should be analyzed, including:

- Examining the current injury rates and severities, comparing them to the rates prior to program implementation (using the goals and objectives as benchmarks).
- Measuring the change in knowledge, behavior, and performance.
- Analyzing the changes in the physical environment.
- Measuring the response to policy changes.

Much of the information needed can come directly from the injury statistics; however, there will be other required measurements. For example, when analyzing the changes in attitude and knowledge, a more **cognitive** measurement may have to be used. Outcome evaluation is often used to measure the results obtained against the overall program goal.

The safety and health program evaluation should include both of these evaluation methodologies. For the purpose of this textbook, these examples are simplistic; however, the reality is that the complete safety and health program must be evaluated. This evaluation can focus on the whole or focus on the individual parts, as long as the evaluation results in a comprehensive look at the program.

cognitive | Skills learned through a mental learning process as opposed to practical learning.

RESPONSIBILITY FOR EVALUATION

Who has the responsibility for program evaluation? The simple answer is that everyone has some responsibility for it. Everyone can and should play some role in the evaluation process. The fire chief or EMS director has the ultimate responsibility to see that program evaluation is done, but the safety program manager may oversee the process. The responders perform a process evaluation each time they respond to an alarm. The evaluation may be accomplished as well through the safety committee, incident safety officers, senior staff members, and first line supervisors.

The safety committee can assist with compiling and analyzing the injury data and can also help with the determination, through feedback, of knowledge, performance, and behavior changes. The incident safety officers are valuable in providing feedback during process evaluation and for measuring compliance with new policies or procedures. Senior management is part of the process by making resources and data available to the safety and health program manager and the safety committee. First-line supervisors also are feedback providers and can very often measure knowledge and attitude changes. First-line supervisors also are integral for feedback on policies and procedure effectiveness. Finally, the individual member should be empowered to make suggestions to improve program components and can measure peer attitude and knowledge changes.

Each of the program evaluators discussed so far has been internal to the organization; however, the *NFPA 1500* standard recommends external evaluations every 3 years.

Having an external evaluation performed is a positive effort for both the safety program manager and the program in general and is highly recommended. Very often a person from the outside who is not close to the issues will find areas in which improvements can be made that may not be as obvious to the internal evaluators. Performing an external evaluation does not have to be a costly endeavor. Because they also have a stake in the safety program, workers' compensation and other insurers may have risk managers on staff and often make these experts available to a client for an external look at the safety program. Personnel such as safety program managers in other local departments may be willing to evaluate the program, as might other fire/EMS chiefs. Local colleges or universities may have a safety and health curriculum and may allow student interns to review your program as a class assignment. Finally, if you want to spend some money, the organization could hire a consultant in the safety and health profession to perform the evaluation.

EVALUATION FREQUENCY

When the program should be evaluated is another question. As mentioned previously, the two types of program evaluation, process and outcome, can help to answer this question. Process evaluation can almost be a continuous process. Outcome evaluation should occur at the end of a specific period of time, usually one year or whatever the program goal or objective defines as a measured time frame.

The frequency of evaluation should be considered dynamic. For example, if a serious injury, near miss, or a death occurs, an immediate evaluation or analysis into the safety program components dealing with the cause should be undertaken. Findings during a PIA may also result in the need to evaluate the program. If new technology is introduced into the profession, an analysis of its applicability to the program should be examined. If there are changes to applicable regulations or standards, the program must be evaluated for compliance.

Program evaluation is a required component of the safety and health program. In fact, program evaluation occurs even before the program starts, because risk identification and analysis are actually types of evaluations. They just happen to be evaluations of a program before a program begins.

SUMMARY

There are two basic formats for the evaluation of the program, the process evaluation and the outcome evaluation. The process evaluation should be viewed as an ongoing analysis into the program to determine whether and to what degree the program is reaching the intended recipients. It also should reveal what changes have occurred in knowledge, performance, and behavior. The outcome evaluation should occur after the program has been in effect for some period of time and actually measures results. These results can be compared to the program goals and objectives and the program effectiveness can be determined.

Everyone in the organization has some role in the evaluation process. Although the responsibility rests with the safety and health program manager, the safety committee, incident safety officers, senior staff, first-line supervisors, and the members all have a role in the evaluation process. External evaluations are recommended in order to have the program evaluated from a different perspective.

Although a process evaluation is ongoing and the outcome evaluation takes place after a defined period of time, the frequency of evaluation is dynamic. If a major event occurs in the organization or if new technologies are introduced, a program evaluation is warranted at that time. After the evaluation, regardless of the type or the frequency, the information gained should be used for program enhancement and improvement.

Concluding Thought: A safety and health program cannot be successful without evaluation methods as part of the process.

REVIEW QUESTIONS

1. List the two types of program evaluations.
2. List the components required for each of the two program evaluations listed in Question 1.
3. List three purposes for program evaluation.
4. What is the role of the first-line supervisor in program evaluation?
5. Describe considerations for determining the frequency for evaluation.

ACTIVITIES

1. Compare the evaluation of the safety program in your organization to that presented in the text. How does your process compare?
2. What changes do you recommend?
3. Are the roles of the various members of your organization similar to the recommendation of the text in terms of program evaluation? If so, are they effective? If not, why are they not effective?

CHAPTER

12
Information Management

LEARNING OBJECTIVES

Upon completion of this chapter, you should be able to:

- Describe the purpose of data collection and reporting.
- Identify the data that should be collected within the organization.
- Identify the data that should be collected for outside organizations.
- Describe the purpose and process for publishing an internal health and safety report.
- Describe the use of the Internet as a health and safety information source.

In May 2009, a fire fighter was seriously injured while operating in a non-designated collapse zone of a commercial structure when an overhang of a bowstring truss roof system collapsed and struck him. The first arriving company officer reported a working fire in a single-story warehouse. The officer looked under a steel roll-up door that was raised approximately three feet off of the ground and saw heavy fire toward the rear of the structure from floor to ceiling. Per department procedures, the first arriving companies went into a "Fast Attack" mode. Crews attempted but were unable to enter the structure because the steel roll-up door wasn't functioning and the main door was heavily secured. The department's Deputy Chief arrived on the scene 9 minutes after the initial crew and determined that the fire should be fought defensively; however, this command was not relayed over the radio or verified with all crews. A crew was operating a 2½-inch hand line just outside the structure, approximately 20 minutes after the first apparatus arrived, when the overhang collapsed and trapped the nozzleman. Key contributing factors identified in this investigation include: scene management and risk analysis, a well-involved fire in a structure with hazardous construction features, and fire fighters operating within a potential collapse area.

Recommendations:

■ Ensure that there are consistent policies and training on an incident management system.

■ Develop, implement, and enforce written standard operating procedures (SOPs) that identify incident management training standards and requirements for members expected to serve in command roles.

■ Ensure that the incident commander conducts an initial size-up and risk assessment of the incident scene before beginning fire fighting operations.

■ Ensure that the first due company officer establishes a stationary command post, maintains the role of director of fireground operations, and does not become involved in firefighting efforts.

■ Implement and enforce written standard operating procedures (SOPs) that define a defensive strategy.

■ Ensure that policies are followed to establish and monitor a collapse zone when conditions indicate the potential for structural collapse.

■ Train all fire fighting personnel on building construction and the risks and hazards related to structural collapse.

■ Conduct preincident planning inspections of buildings within their jurisdictions to facilitate development of safe fireground strategies and tactics.

The use of information technology is a tool for the safety program manager to maintain awareness of incidents that occur throughout the emergency service. This information can help guide the risk assessment process and provide excellent ideas and comparative data as safety and health program development progresses.

Source: Information for this case study was taken in part from CDC/ NIOSH Fire Fighter Fatality Investigation Report F2009-21. http://www .cdc.gov/niosh/fire/reports/face201021.html

INTRODUCTION

Like program evaluation, data collection and information analysis are critical components of the safety and health program. Data collection and analysis are used for many of the other program components discussed previously, including the development of goals and objectives and the evaluation of the program or its components.

NOTE In order to be useful, data must be retrievable in a format that is compatible with its intended use.

This process should be facilitated through the use of a computer for easy analysis and retrieval.

Data is collected for use within the organization. Data is also collected for and by outside agencies and used for various other purposes. The collection and analyzation of data by outside agencies is helpful for local comparison to national trends. Once the internal data is collected and analyzed, an annual health and safety report should be published for all levels of the organization.

The environment is rapidly changing and so is the speed with which we can communicate. Our access to information has never been greater. The safety and health program manager can use the Internet to obtain volumes of information regarding safety and health regulations and standards, as well as copies of policies from other similar departments regarding their safety and health programs.

The following Firefighter Life Safety Initiatives relate to the use of information management to improve responder safety:

    7. Create a national research agenda and data collection system that relates to the initiatives.

    8. Utilize available technology wherever it can produce higher levels of health and safety.

PURPOSE OF DATA COLLECTION AND REPORTING

There are many reasons for performing data collection. For starters, data collection and reporting is required for many jurisdictions by OSHA or a state department of labor. It is equally important for the organization to have the information available for analysis and evaluation. Furthermore, good record keeping helps to protect the organization and the members from a legal standpoint (see **Figure 12-1**).

Records regarding workers' compensation and other insurance claims can help predict increases or decreases in premiums that will have future budget impacts. Further, the mandatory reporting requirements of federal and state agencies usually carry fines associated with noncompliance.

Bear in mind that most, if not all, medical-related data and information is considered confidential, and access to the information should be kept to only those with a need to know. The first-ever federal privacy standards to protect patients' medical records and other health information provided to health plans, doctors, hospitals, and other health care providers took effect on April 14, 2003. This standard provided patient privacy protections as part of the Health Insurance Portability and Accountability Act of 1996 (HIPAA). This act probably will have some impact on the medical data collected as part of the safety and health program, depending on the type of organization. Therefore, it is important during the development and evaluation of the safety and health program that legal advice is obtained regarding injury reporting and privacy issues. After seeking legal direction, department SOPs must be developed and implemented that address who has access to medical-related data and why.

INTERNAL DATA COLLECTION

Internal data collection is associated with health and safety data generated from within the organization and is used within the organization. Several common data types fall into this category, including:

- Injury reports
- Accident reports

SOUTHWEST FLORIDA PROFESSIONAL FIREFIGHTERS
LOCAL 1826 / I.A.F.F.
EXPOSURE REPORT

1. Firefighters Name _____ 2. Social Security # _____ 3. 1826 Dist. # _____

4. Incident Date _____ 5. Alarm Time _____ 6. Incident # _____

A MEDICAL / RESCUE EXPOSURE

7. Patients Name _____

8. Facility Patient Transported To _____

9. Transporting Agency _____

10. Name of Suspected or Known Disease: _____

11. Precautions Taken By You at Scene and After Exposure

☐ gloves ☐ face mask ☐ goggles ☐ protective gown

☐ washed exposed skin ☐ physician treatment (if this is checked complete *section F*)

	How you were exposed	A. Topical Skin Contact	B. Needle Puncture	C. Direct contact to open wound	D. In Eyes	E. In nose or mouth
12.	BLOOD					
13.	URINE					
14.	SPUTUM					
15.	VOMITUS					
16.	FECES					
17.	OTHER					

B FIRE INCIDENT TYPE

18. ☐ Residential Fire 19. ☐ Industrial Fire 20. ☐ Vehicle Fire 21. ☐ Commercial Fire 22. ☐ Wildland Fire 23. ☐ Trash/Dumpster fire 24. ☐ Marine Fire 25. ☐ Explosion 26. ☐ Rescue 27. ☐ Spill

28. Other (describe in 1 or 2 words) _____ 29. Type of occupancy (1 or 2 words) _____

C LENGTH OF EXPOSURE BY FIRE STAGE / ACTIVITY

FIRE STAGE :	A	B	C	D
	<1 hr.	1-2 hr	2-3 hr	3 + hrs
30. Incipient				
31. Free Burning				
32. Smoldering				
33. Non-Fire Incident				

	A	B	C	D
	< 1 hr	1-2 hr	2-3 hr	3 + hrs
34. Extinguishment				
35. Empty / Ventilation				
36. Rescue				
36. Light Overhaul				
37. Heavy overhaul				
38. E.M.S.				
39. Investigation				

D SMOKE / CHEMICAL/ OTHER EXPOSURE

41. Smoke Conditions ☐ light ☐ Heavy ☐ None

42. Smoke Color _____

Chemicals Present	A Vapor Gas	B Dust	C Liquid Heavy Mist	D Light Mist	E Smoke	F Solid Powder
43.						
44.						
45.						
46.						

47. Type of Chemical Exposure to Above: ☐ Inhaled ☐ Ingested ☐ Skin Contact
   ☐ Eye Contact ☐ Other _____

E SYMPTONS

	AT INCIDENT	SYMPTON	AFTER INCIDENT	FOR HOW LONG
48.		Eyes Burn		
49.		Cough		
50.		Nose Bleed / Cough Blood		
51.		Nose/Lung Irritation		
52.		Nausea/Queasiness		
53.		Dizzy		
54.		Ears Ringing		
55.		Headache		
56.		Skin Rash/Burn		
57.		Unconscious		
58.		Other:		

F MEDICAL DIAGNOSIS

59. DID YOU RECEIVE MEDICAL EVALUATION OR TREATMENT AFTER EXPOSURE? ☐ YES ☐ NO

60. DIAGNOSIS _____

61. DOCTORS NAME _____ 62. FACILITY _____

G SPECIAL EQUIPMENT / DECONTAMINATION ☐ PASS ALARMS ☐ BUNKER PANTS ☐ HELMETS ☐ BUNKER COATS ☐ GLOVES ☐ HOODS ☐ SCBA'S ☐ BOOTS

63. WERE YOU SUPPLIED SPECIAL EQUIPMENT FOR THIS INCIDENT? ☐ YES ☐ NO DESCRIBE _____

64. WERE SPECIAL DECONTAMINATION PROCEDURES FOLLOWED AFTER THE EXPOSURE ☐ YES ☐ NO DESCRIBE _____

H CO-WORKERS AT TIME OF EXPOSURE

65. PLEASE LIST NAMES OF OTHER FIREFIGHTERS WORKING CLOSE TO YOU AT TIME OF EXPOSURE _____

I ADDITIONAL INFORMATION (YOUR FUNCTION AT INCIDENT, UNUSUAL CIRCUMSTANCES, ETC.)

66. _____

J 67. DATE REPORT FILLED OUT _____ 68. NAME OF PERSON FILLING OUT REPORT (PRINT) _____

69. SIGNATURE _____

FIGURE 12-1

An example of an exposure-reporting form.

- Individual medical histories
- Drug-free workplace test results
- Reports dealing with an employee killed in the line of duty
- Exposure records

In order to collect the data required for these reports, standardized forms are often used. The data should then be placed into a computer database for storage and later retrieval. Safety and health program managers should check their local, state, and federal laws for requirements for keeping hard copies of these records and for the length of time they must be maintained. Regardless of the hard-copy requirements, the information should still be filed electronically—in addition to the required hard copies —for easy retrieval and analysis. **Figure 12-2** is an example of an accident/injury reporting form with the generally required information.

EXTERNAL DATA COLLECTION

External data is the information, collected internally, that is to be used by another agency outside of the organization. This outside agency could be state and national databases, workers' compensation carriers, or insurance companies. As might be recalled from Chapter 1, the IAFF/IAFC Joint Labor Management Wellness/Fitness Initiative requires reporting to a national database accessible by local organizations.

The information required for the external agencies may or may not be the same as that needed within the organization. However, the internal use forms usually provide adequate background information for the external reporting requirements.

Workers' Compensation

Clearly, the organization's workers' compensation carrier is one external organization that wants injury data. Generally, there are rules and laws that provide for specific time frames in which a workers' compensation carrier must be notified. The workers' compensation carrier will also, upon request, provide the organization with a summary of claims over a requested period of time. This can be a useful tool in program evaluation. For example, knowing that internally the number of workers' compensation claims have gone down is good news. However, if the costs of the claims, in terms of lost time and treatment, have increased, the program will have to be reevaluated and changes implemented.

Occupational Safety and Health Administration

Although OSHA does not apply to every public fire department in the country, there are reporting requirements in the states in which OSHA does apply. The OSHA 1904

Report #_____

Accident / Injury / Investigation Report
PRINT ONLY

TO BE FILLED OUT BY INJURED PARTY   **Today's Date:**_____

Name of Injured:	Sex: Male ❑ Female ❑	Age:

Date & Time of Accident: AM   PM | Location of Accident:

Describe the accident and how it occurred:

Was accident reported to immediate supervisor? Yes ❑ No ❑ | Date & Time reported to Supervisor:

Were there any injuries?   Yes ❑ No ❑ If yes, describe injuries:

Was medical treatment given? Yes ❑ No ❑   If yes, give physician name & address:

Was time lost from work? Yes ❑ No ❑   If yes, explain.

Were department SOP's adhered to? Yes ❑ No ❑   If no, explain.

Were safety standards adhered to? Yes ❑ No ❑   If no, explain:

Was law enforcement notified? Yes ❑ No ❑   If yes, which law enforcement agency?

What could have been done to avoid this accident?

SIGNATURE OF INJURED PERSON:

TO BE FILLED OUT BY IMMEDIATE SUPERVISOR

Cause of accident:

Witnesses to accident? Yes ❑ No ❑   If yes, list witnesses.

Was personal protective equipment required? Yes ❑ No ❑   Was it being used? Yes ❑ No ❑   If NO, explain:

Interim corrective actions taken to prevent recurrence:

Permanent corrective action recommended to prevent recurrence:

SUPERVISOR SIGNATURE   **DATE:**

Status and follow-up action taken by Safety Officer

SAFETY OFFICER SIGNATURE   **DATE:**   **REVIEWED BY CHIEF / DATE**

Courtesy of Palm Harbor Fire Rescue

FIGURE 12-2
Sample accident/injury-reporting form.

regulation requires that the **OSHA 300** log (see **Figure 12-3**) and **OSHA 301** incident report be completed within seven (7) calendar days of receiving information that a recordable injury or illness has occurred. All of the entries on the OSHA 300 log must be summarized on the **OSHA 300-A** log at the end of the year. OSHA regulations do provide for the privacy of the injured party. The safety and health program manager should refer to OSHA 1904 for these provisions.

However, there are guidelines on what types of injuries must be reported. An injury must be reported if it results in any of the following: death, days away from work, restricted work or transfer to another job, medical treatment beyond first aid, or loss of consciousness. You must also consider a case to meet the general recording criteria if it involves a significant injury or illness diagnosed by a physician or other licensed health care professional, even if it does not result in death, days away from work, restricted work or job transfer, medical treatment beyond first aid, or loss of consciousness.

Organizations that do not fall under OSHA for public emergency response agencies should contact their state department of labor to determine a state requirement for similar reporting.

OSHA 300	Log of work-related injuries and illnesses.
OSHA 301	The injury and illness incident report.
OSHA 300-A	Summary of work-related injuries and illnesses.

National Fire Protection Association

As described in Chapter 1, the NFPA publishes an annual injury report and an annual firefighter fatality report. In order for these reports to reflect a national experience, organizations are asked to participate in the NFPA data collection program. Remember that the injury portion of the NFPA's reports are a sampling designed to predict the national experience.

United States Fire Administration

The USFA also does health and safety data collection through the National Fire Incident Reporting System (NFIRS). Although voluntary, the NFIRS has components for reporting firefighter casualties, both fatalities and injuries. This information is reduced to an annual report that addresses the nationwide experience for those organizations that participate.

OSHA's Form 300 (Rev. 01/2004)

Log of Work-Related Injuries and Illnesses

You must record information about every work-related death and about every work-related injury or illness that involves loss of consciousness, restricted work activity or job transfer, days away from work, or medical treatment beyond first aid. You must also record significant work-related injuries and illnesses that are diagnosed by a physician or licensed health care professional. You must also record work-related injuries and illnesses that meet any of the specific recording criteria listed in 29 CFR Part 1904.8 through 1904.12. Feel free to use two lines for a single case if you need to. You must complete an Injury and Illness Incident Report (OSHA Form 301) or equivalent form for each injury or illness recorded on this form. If you're not sure whether a case is recordable, call your local OSHA office for help.

Identify the person			Describe the case		
(A) Case no.	(B) Employee's name	(C) Job title (e.g., Welder)	(D) Date of injury or onset of illness	(E) Where the event occurred (e.g., Loading dock north end)	(F) Describe injury or illness, parts of body affected, and object/substance that directly injured or made person ill (e.g., Second degree burns on right forearm from acetylene torch)

Public reporting burden for this collection of information is estimated to average 14 minutes per response, including time to review the instructions, search and gather the data needed, and complete and review the collection of information. Persons are not required to respond to the collection of information unless it displays a currently valid OMB control number. If you have any comments about these estimates or any other aspects of this data collection, contact: US Department of Labor, OSHA Office of Statistical Analysis, Room N-3644, 200 Constitution Avenue, NW, Washington, DC 20210. Do not send the completed forms to this office.

Source: U.S Department of Labor, Occupational Safety & Health Administration

FIGURE 12-3
OSHA 300 Log.

International Association of Fire Fighters

Like the NFPA and OSHA, the IAFF collects data for annual injury, exposure, and fatalities reports. The data collected by the IAFF is only from paid fire departments with IAFF affiliation and includes the United States and Canada. The published report considers EMS-related activities to a greater degree than the previous two reports.

Attention: This form contains information relating to employee health and must be used in a manner that protects the confidentiality of employees to the extent possible while the information is being used for occupational safety and health purposes.

Year 20__ __

U.S. Department of Labor
Occupational Safety and Health Administration

Form approved OMB no. 1218-0176

Establishment name _____

City _____ State _____

Classify the case CHECK ONLY ONE box for each case based on the most serious outcome for that case:				**Enter the number of days the injured or ill worker was:**		**Check the "Injury" column or choose one type of illness:**					
		Remained at Work				(M)					
Death	Days away from work	Job transfer or restriction	Other recordable cases	Away from work	On job transfer or restriction	Injury	Skin disorder	Respiratory condition	Poisoning	Hearing loss	All other illnesses
(G)	(H)	(I)	(J)	(K)	(L)	(1)	(2)	(3)	(4)	(5)	(6)
☐	☐	☐	☐	____ days ____ days		☐	☐	☐	☐	☐	☐
☐	☐	☐	☐	____ days ____ days		☐	☐	☐	☐	☐	☐
☐	☐	☐	☐	____ days ____ days		☐	☐	☐	☐	☐	☐
☐	☐	☐	☐	____ days ____ days		☐	☐	☐	☐	☐	☐
☐	☐	☐	☐	____ days ____ days		☐	☐	☐	☐	☐	☐
☐	☐	☐	☐	____ days ____ days		☐	☐	☐	☐	☐	☐
☐	☐	☐	☐	____ days ____ days		☐	☐	☐	☐	☐	☐
☐	☐	☐	☐	____ days ____ days		☐	☐	☐	☐	☐	☐
☐	☐	☐	☐	____ days ____ days		☐	☐	☐	☐	☐	☐
☐	☐	☐	☐	____ days ____ days		☐	☐	☐	☐	☐	☐
☐	☐	☐	☐	____ days ____ days		☐	☐	☐	☐	☐	☐
☐	☐	☐	☐	____ days ____ days		☐	☐	☐	☐	☐	☐
☐	☐	☐	☐	____ days ____ days		☐	☐	☐	☐	☐	☐

Page totals ▶ ___ ___ ___ ___

Be sure to transfer these totals to the Summary page (Form 300A) before you post it.

						Injury	Skin disorder	Respiratory condition	Poisoning	Hearing loss	All other illnesses
				Page ___ of ___		(1)	(2)	(3)	(4)	(5)	(6)

Source: U.S Department of Labor, Occupational Safety & Health Administration

National Institute for Occupational Safety and Health

NIOSH began a project in 1997 in which it investigates firefighter line-of-duty deaths. The project, called the Fire Fighter Fatality Investigation and Prevention Program, was funded in 1998. **Text Box 12-1** describes this program, while **Figure 12-4** shows the NIOSH decision flow chart.

TEXT BOX 12-1

DESCRIPTION OF THE NIOSH FIRE FIGHTER FATALITY INVESTIGATION AND PREVENTION PROGRAM

Fire Fighter Fatality Investigations

The NIOSH Fire Fighter Fatality Investigation and Prevention Program (FFFIPP) conducts investigations of fire fighter line-of-duty deaths to formulate recommendations for preventing future deaths and injuries. The FFFIPP is a public health practices investigation program. NIOSH investigators are not conducting their investigation to enforce compliance with state or federal occupational safety and health standards and do not determine fault or place blame on fire departments or individual firefighters. The program's goal is to learn from these tragic events and prevent future similar events through its recommendations. NIOSH does not investigate every fire fighter fatality. The FFFIPP has investigated approximately 40% of fire fighter deaths since the program's start in 1998.

Investigations are prioritized based on a consideration of a series of factors, as outlined in the program's decision flow chart available at http://www.cdc.gov/niosh/fire/ (the FFFIPP's website). Investigation priorities may change based on a review of fatality data on leading risks to fire fighters as well as stakeholder input. NIOSH conducts investigations of both career and volunteer fire fighter line-of-duty deaths. NIOSH is notified of a line-of-duty death in a number of ways, including notification by the United States Fire Administration (USFA), a fire department representative, the International Association of Fire Fighters (IAFF), and state fire marshal's offices. NIOSH may also learn of fatal events through media coverage before being notified by other agencies. NIOSH reviews each event and makes a determination as to whether or not to initiate an investigation. NIOSH may also investigate selected injury and near-miss events that represent the potential for new or emerging hazards to the fire service. A fire department's decision to participate in the FFFIPP is voluntary. Participating fire departments and fire fighters recognize the value of an objective independent investigation process that focuses on developing recommendations to prevent future fire fighter injuries and deaths.

Program Objectives

- Better identify and define the characteristics of line-of-duty deaths among fire fighters.

(continues)

- Develop recommendations for the prevention of deaths and injuries.
- Disseminate prevention strategies to the fire service.

Traumatic Injury Deaths

The program conducts investigations of fireground and non-fireground fatal injuries resulting from a variety of circumstances, such as motor vehicle incidents, burns, falls, structural collapse, diving incidents, and electrocutions. NIOSH staff also conduct investigations of selected non-fatal injury events that suggest the potential to identify new or emerging hazards. NIOSH staff with expertise in respirators and fire fighter protective clothing assist with investigations of incidents in which the function of respiratory protective equipment or protective clothing may have been a factor. Each investigation results in a report summarizing the incident and includes recommendations for preventing future similar events.

Cardiovascular Disease (CVD) Deaths

NFPA data show that sudden cardiac death is the most common type of on-duty death for fire fighters. NIOSH investigations of these fatalities include assessing the contribution of personal and workplace factors. Personal factors include identifying individual risk factors for coronary artery disease. The workplace evaluation includes the following assessments:

- Estimating the acute physical demands placed upon the fire fighter.
- Estimating the fire fighter's acute exposure to hazardous chemicals.
- Assessing efforts by the fire department to screen for coronary artery disease.
- Assessing efforts by the fire department to develop fitness and wellness programs.

Information Dissemination

FFFIPP posts all investigative reports onto the NIOSH website and notifies subscribers to the FF Safety Announcements of each posting. In addition to website postings, selected reports are packaged and mailed periodically to fire departments and emergency medical services organizations across the country. NIOSH reports and publications are also distributed at major fire service conferences, training centers,

(continues)

and other venues. All NIOSH reports and publications are public domain information and may be freely copied and reproduced for training and educational purposes.

What to Expect During a NIOSH Investigation

Once notified of a fatality, NIOSH makes a determination as to whether to conduct an investigation. Factors that are considered include the type of event, whether that type of event is a priority for investigation, and investigator workload (a decision flowchart is available on the FFFIPP website). If the decision is made to conduct an investigation, a NIOSH representative will contact the fire department to enlist their cooperation and schedule a site visit.

For traumatic injury deaths, NIOSH seeks to conduct a site visit within three weeks of the incident, and NIOSH investigators visit the incident site to gather information and take pictures and measurements. NIOSH investigators will review all applicable documents (e.g., department standard operating procedures, dispatch records, training records for the victim, Incident Commander and officers' reports, the victim's medical records (where applicable), coroner/medical examiner's reports, death certificates, blueprints of the structure, police reports, photographs, and video). In incidents suggestive of potential performance problems with respirators or personal protective clothing, investigators will request that the equipment or clothing be sent to the NIOSH National Personal Protective Technology Laboratory for evaluation. Additionally, investigators will interview fire department personnel and fire fighters who were on the scene at the time of the incident. The interview process is voluntary and witnesses' statements are not made under oath or reviewed by the witness. Because the interviews are not recorded, NIOSH investigators rely on their own notes from these interviews and the applicable documents to develop a description of the conditions and circumstances leading to the fatalities or injuries. NIOSH provides these descriptions in their reports solely for the purposes of providing context for the agency's recommendations for preventing similar occurrences. NIOSH may work closely with other investigating agencies. When NIOSH does not have the necessary subject matter expertise, NIOSH will enlist the assistance of other experts, such as experts in motor vehicle incident reconstruction or fire growth modeling.

(continues)

Once the investigation is completed, NIOSH will summarize the sequence of events related to the incident and prepare a draft report. Each department, union (if present), or family (where applicable, because some draft reports includes personal or medical history) will have the opportunity to review this portion of the report in draft form to help ensure it is factually accurate. The report is then finalized with the addition of recommendations for preventing deaths and injuries under similar circumstances. In traumatic injury incidents, NIOSH will enlist the assistance of fire service subject-matter experts to review the draft reports and recommendations. Once the fire department, union (if present), and family (where applicable) have received the final copy of the NIOSH incident report, it is made available to the public through the Fire Fighter Fatality Investigation and Prevention Program website. The reports are anonymous and do not name the fire department, victim(s), or other fire fighters involved in the incident.

Other Publications and Products

In addition to the incident reports developed by the program, NIOSH has produced other products related to fire fighter safety and health. A list of these products is available on the website described previously.

Source: NIOSH

PUBLISHING THE HEALTH AND SAFETY REPORT

Once a data collection and retrieval system is put into place, the measurable data must be turned into usable information. Normally, this requires some sort of analysis and the compiling of a report over time, usually annually (see **Figure 12-5**). Once published, this report should be distributed to all levels of the organization, including elected officials and senior city or county officials. This report can help as an educational tool when support for resources is needed to enhance the program. The annual health and safety report should contain the following information (keeping in mind, again, the confidentiality of medical records):

- Introduction
- General state of the organization in terms of safety and health
- Accomplishments/improvements/benchmarks of the safety and health program for the reporting time period
- Goals and objectives for the next reporting period
- The analysis of the injury and fatalities for the reporting year

FIGURE 12-4
NIOSH decision flow chart.

FIGURE 12-5
Annual activity reports should include safety and health information.

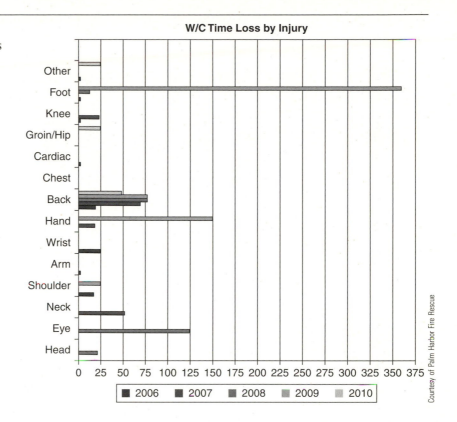

W/C Time Loss by Injury

Courtesy of Palm Harbor Fire Rescue

- A comparison to the national experience
- A comparison to similar departments in similar geographic areas
- A report on significant incidents, including findings and changes made for preventing future occurrences
- A summary of the organization's compliance with regulations and standards
- Other plans for improvement and requested resources
- A summary
- Specific graphics relating to the presentation of the information

Sometimes the health and safety officer will be called on to officially present this report to the senior management or the elected officials. In such cases, the presenter should have adequate audiovisual aids to assist in this process, which might include slides, overheads, or computer-generated graphics, charts, bulleted goal and objective lists, and sometimes pictures.

ACCESSING HEALTH AND SAFETY INFORMATION USING THE INTERNET

With electronic mail fax, social media, instant messaging, and video conferencing, communication is instantaneous. Access to information continues to grow. The safety and health program manager, through the use of a computer, can have access to a great deal of information that can assist with the safety and health program. Through the use of an organization's e-mail system, accident and injury reports can be filed on a timelier basis, and policy changes can be distributed to all work locations with the touch of a computer key. Granted, not all organizations have access to some of these systems, but for those that do, the amount and the quality of the information is invaluable.

NOTE The Internet can provide access to other organizations' information, state and federal regulations, and annual national reports such as those available through the USFA.

Internet search engines can be used to search for documents or web pages that would have the words or phrases being searched for. In the following list, several popular sites for safety and health information are described. Each also commonly has **links** to other pages with similar information. The reader is encouraged to visit these sites and to search for others. The information gained is timely and useful to the safety program. However, a note of caution: Information on the Internet is not guaranteed to be factual. If there is any doubt, the user should verify the validity of the information. Often the search produces so many **hits** that it would be unreasonable to look at all of them. In this case, the user may want to be more specific to the subject they are researching. Some examples or common sites of interest to the safety and health manager are described in **Text Box 12-2.**

Internet search engines Programs on the Internet that allow a user to search the entire Internet for key words or phrases.

links Used on Internet pages so that each page may be directly tied to a page on another Internet site.

hits Number of documents found when searching for a key word or phrase using a search engine.

TEXT BOX 12-2

SOME EXAMPLE SITES THAT SHOULD BE VISITED BY THE SAFETY AND HEALTH MANAGER

- The Department of Labor has a website that gives access to OSHA regulations and a history of organizations that were cited for specific violations. http://www.osha.gov

- The International Association of Fire Chiefs and the International Association of Fire Fighters each have websites that can provide usable information. http://www.iafc.org and www.iaff.org

- DHS, FEMA, the USFA, and the National Fire Academy have a very comprehensive site with a great deal of health and safety information and advice on where to find resource information. http://www.usfa.fema.gov

- The Centers for Disease Control and Prevention site can be helpful in obtaining information regarding infectious diseases and information on how to comply with regulations dealing with infectious diseases. This site is also where NIOSH is hosted. http://www.cdc.gov and http://www.cdc.gov/niosh/fire/

- The NFPA provides many of the safety and health standards. http://www.nfpa.org

- There are numerous sites for fire/EMS magazines and journals.

- Some worker's compensation carriers are allowing for Internet access to an organization's claim records. The access is limited to those who need the information for tracking claims or who are responsible for the health and safety program.

© Cengage Learning 2013

SUMMARY

Data collection and information analysis are important components of the safety and health program. The data collection needs will vary based on whether the information is for use within the department or for an external agency. Each can be turned into valuable feedback that will assist in the evaluation of the program. Regardless of where the information is to be used, a form or similar consistent means should be established for gathering the data. On scene, information can begin to accumulate.

Once analyzed, the information should be compiled into an annual report for organization-wide distribution. The annual report has a number of required sections and should answer these questions: Where were we? Where are we now? Where are we going? And, how can we get there?

By using computers and the Internet, an organization can communicate much faster and can access much more information more easily than has ever been possible.

Many national, state, and local health and safety–related agencies have information posted on the Internet, as do local emergency response organizations. This information can be a valuable tool in assisting the health and safety program manager and safety committee in program design, training, and evaluation.

Concluding Thought: As with the other components of the program, information management is essential. Through the use of computers and the Internet, there is virtually no limit to the access of this information and new information. ICS and OSHA logs are good starting points.

REVIEW QUESTIONS

1. List three reasons for data collection and reporting.
2. Compare data collection from the internal perspective to the external perspective.
3. Injury data downloaded from the USFA website would be considered internal data.
   a. True
   b. False
4. Which of the following are essential to an organization's information system?
   a. Data must be collected using a common means.
   b. Data must be stored so that it is easily retrievable.
   c. Data must be analyzed and put in a useful format.
   d. All of the above are essential components.
5. List the sections that should be a part of the annual safety and health report.

ACTIVITIES

1. Compare the annual health and safety report from your department with the recommendations in this text. Are all the sections included? Is the document used as a planning tool?
2. Analyze the procedures in your department for collecting and storing health and safety data. Is the data retrievable in a usable format? Is the data useful as information?
3. Access the Internet and visit the sites that are recommended in the text and search for others. What information can be useful in your department?

CHAPTER 13

Special Topics in Emergency Service Occupational Safety and Health

LEARNING OBJECTIVES

Upon completion of this chapter, you should be able to:

- Describe the legal considerations of the safety and health program.
- Describe ethical considerations of the safety and health program.
- Describe financial considerations of the safety and health program.
- Describe health and safety considerations in the diversified workforce.
- Describe the safety and health program implications of future trends and technologies.

On June 30, 2011, a fire fighter (FF) responded to a residential structure fire. The FF, wearing full turnout gear and self-contained breathing apparatus (SCBA) on-air, climbed a 14-foot ladder to the second floor and performed exterior and interior fire suppression activities for about 30 minutes. After the fire was brought under control, he started to perform overhaul (mop-up) operations on the second floor when he suddenly collapsed. Crew members carried the FF down the stairs and outside the dwelling. Cardiopulmonary resuscitation (CPR) was begun as an ambulance was requested. The ambulance arrived, advanced life support was provided, and the FF was transported to the local hospital's emergency department (ED). In the ED, the FF regained a pulse but remained minimally responsive and was flown to a regional hospital where angioplasty was performed. Over the next 8 days, his condition did not improve and, after consulting with the family, the decision was made to remove the FF from life support. The attending physician pronounced him dead at 0434 hours on July 8, 2011. The death certificate listed "myocardial infarct" due to "cardiac arrest" as the cause of death. No autopsy was performed. Carboxyhemoglobin levels were not measured to test for carbon monoxide exposure. Given the FF's known underlying coronary artery disease (CAD), NIOSH investigators concluded that the physical exertion of responding to the call and suppressing the fire triggered his heart attack, which caused his death.

NIOSH investigators offer the following recommendations to address general safety and health issues. Had some of these recommended programs been in place, it is possible the FF's death may have been prevented:

- Provide preplacement and annual medical evaluations to all fire fighters.

- Ensure fire fighters are cleared for return to duty by a physician knowledgeable about the physical demands of fire fighting, the personal protective equipment used by fire fighters, and the various components of *National Fire Protection Association (NFPA) 1582*.

- Phase in a comprehensive wellness and fitness program for fire fighters.

- Perform a preplacement and annual physical performance (physical ability) evaluations.

- Provide fire fighters with medical clearance to wear self-contained breathing apparatus (SCBA) as part of the Fire Department's medical evaluation program.

- Use a secondary (technological) test to confirm appropriate placement of the endotracheal tube.

- Perform an autopsy on all on-duty fire fighter fatalities.

The organization's health and safety officer must maintain awareness of new technologies and best practices for implementation into operations. They also must ensure proper funding for the safety and health program and be aware of emerging trends.

Source: Information for this case study was taken in part from CDC/ NIOSH Fire Fighter Fatality Investigation Report F2011-16. http://www .cdc.gov/niosh/fire/reports/face201116.html

INTRODUCTION

This text and study of occupational safety and health would not be complete without a discussion of the legal, ethical, and financial issues that could affect the safety and health program. This chapter discusses these issues as well as how an increasingly diverse workforce and future trends and technologies may affect the safety and health program.

Some of the future trends presented are forward looking and therefore should not be considered complete. Instead, they are designed to spark interest and future research, and to produce changes in the emergency service field in order to be prepared for the challenges that are sure to be ahead.

Grants are one means to fund health and safety programs, as suggested in the following Firefighter Life Safety Initiative:

> 10. *Grant programs should support the implementation of safe practices and/or mandate safe practices as an eligibility requirement.*

LEGAL ISSUES

A number of legal issues affect emergency service organizations.

NOTE Legal action against emergency service organizations was once uncommon; however, there has been a significant increase in these cases.

Some of these legal issues can have a direct impact on or relationship to the safety and health program. For example, the safety program requirement for emergency vehicle operators could be called into question after an apparatus is involved in an accident.

NOTE The safety and health program manager should be aware of some of the more common legal issues related to the safety and health program.

The following snapshot of some of these issues should not be considered all-inclusive, nor should it be considered legal advice—it is simply meant to provide awareness. When specific issues arise, the department's legal counsel must be contacted.

Tort Liability

The concept of liability is important to the safety and health manager as well as to department members. The safety and health program could be questioned should a claim of liability arise. **Tort liability** is an action that harms another person or group and can occur when a person or group of people act, or fail to act, without the right to do so, and thus harm another directly, or indirectly. The following four types of tort liabilities are applicable to the safety and health program:

1. *Strict Liability.* Violation of the law or other regulations, even if violation is unintentional.
2. *Intentional Liability.* Violation of the law or other regulations knowingly, and harm results.
3. *Negligent Liability.* Person fails to do what a reasonable and prudent person would have done under the same or similar circumstances.
4. *Warrant Liability.* Promised service level is not delivered and harm results.

It is easy to see where these types of liabilities can have an impact on the safety and health program. For example, suppose a department is required to follow OSHA 1910.134, Respiratory Protection Regulation, of which SCBA mask fit testing is a component. If the department does not fit test its employees, is it guilty of strict or intentional liability? Maybe. In order to prove liability, four elements must be present:

- Duty or standard to act
- Breach of duty, either an action or omission
- Failure caused the harm
- Actual measurable loss or harm

In the SCBA example, if an employee had been injured as a result of smoke inhalation because of a poor fitting mask, would the department have liability?

Text Box 13-1 gives some examples of ways that liability can arise.

The issue of liability is complex and is changing with every court decision.

NOTE The emergency service administration and safety and health officer must have a close working relationship with the legal staff.

---

tort liability An action that harms another person or group and can occur when a person or group of people act, or fail to act, without the right to do so, and thus harm another directly, or indirectly.

TEXT BOX 13-1

SITUATIONS THAT CAN RESULT IN A LIABILITY CLAIM

- Poor or out-of-date plans
- Personnel not trained; undocumented or unsafe training
- Hazards and risks not identified
- Personnel not warned of hazards
- Equipment not operated or maintained properly
- Standards not present or enforced
- Failure to follow the standard of care (refer to Chapter 2)

© Cengage Learning 2013

American with Disabilities Act

Title I of the Americans with Disabilities Act (ADA) of 1990 took effect July 26, 1992. This act prohibits private employers, state and local governments, employment agencies, and labor unions from discriminating against qualified individuals with disabilities in job application procedures, hiring, firing, advancement, compensation, job training, and other terms, conditions, and privileges of employment. An individual with a disability is a person who has one or more of the following:

- Has a physical or mental impairment that substantially limits one or more major life activities.
- Has a record of such an impairment.
- Is regarded as having such an impairment.

The emergency service safety and health manager should understand the provisions in Title I of the ADA. One important provision is that of medical examinations and disability.

NOTE Specifically, an organization may not ask job applicants about the existence, nature, or severity of a disability.

Applicants may be asked about their ability to perform specific job functions. A job offer may be conditioned on the results of a medical examination, but only if the examination is required for all entering employees in similar jobs.

NOTE Medical examinations of employees must be job related and consistent with the employer's business needs.

Again, it is necessary that the safety and health program comply with this law.

Age Discrimination in Employment Act

The Age Discrimination in Employment Act of 1967 (ADEA) is another federal regulation that should be a consideration in the safety and health program.

NOTE The ADEA protects individuals who are 40 years of age or older from employment discrimination based on age and applies to both employees and job applicants.

Under the ADEA, it is unlawful to discriminate against a person because of age with respect to any term, condition, or privilege of employment including, but not limited to, hiring, firing, promotion, layoff, compensation, benefits, job assignments, and training. The ADEA applies to employers with 20 or more employees, including state and local governments. It also applies to employment agencies, labor organizations, and the federal government.

ETHICAL ISSUES

NOTE Ethics could be defined simply as doing the right thing.

There are ethical considerations in every business or organization. There are also ethical issues for the safety and health manager and others involved in the safety and health program. One safety and health ethical concept is that members of an organization should be provided with the safest work environment possible. In emergency services, in terms of incident response and operation, we seldom have control of the work environment; however, in terms of operations, facilities, and equipment, the organization can do everything possible to ensure the members' safety.

Simply in the course of accident investigation and record management, the members involved directly with the program—including the safety and health officer, management, and the safety committee—are often privy to confidential information. Members of the program that have access to this information must know that specifics of events and other information about coworkers should not be discussed with other members. The trust and respect of confidential information is essential for the success of the program. Members involved in the safety and health program must follow the ethical standards required by the organization; however, they must understand and be educated in further ethical standards that can be applied to them as a result of their relationship to the safety and health program.

FINANCIAL ISSUES

As described in the preceding chapters, components of safety and health programs cost money; however, the cost of not having a comprehensive program is more than the costs associated with program operations. There are some common methods for funding the program. The first and most common method is to use funds from the organization's annual budget. Another is to seek grant funding.

NOTE Often, to get the necessary funds appropriated for a safety and health program, either through the annual budget or through grant funding, a cost-benefit analysis must be performed and presented.

Annual Budget Process

NOTE The most common funding source is an organization's annual budget.

The safety and health officer, as part of the senior staff, should have input into the budget's preparation. The safety and health officer is often in a position of salesperson, selling the benefits of a particular part of the program for which they are seeking funding. A complete cost-benefit analysis is necessary and helpful to persuade funders of the necessity for appropriation of funds.

Aside from new components that the safety and health officer may propose each year, ongoing costs must be considered, including:

- Salary and benefits for the safety and health officer
- Annual medical exams for personnel
- Fitness equipment or program maintenance
- Funds for safety and health training of personnel
- Payment for medical services and department physician
- PPE replacement and maintenance

The safety and health officer has to anticipate and project these costs each year. Most are operating costs from the budget, but there will be capital items as well, such as fitness equipment. There are some opportunities for grant or foundation funding for starting components of the safety and health program, but this grant funding is often limited to capital items.

Grant Funding

NOTE There are a number of funding sources from grants. Some are federal, some state, and even some at the local city or county level.

Sources of grant funding can be researched on the Internet; http://www.grants.gov is one such site. There are also subscriptions to grant newsletters and even consultants that can help a department look for funds and write the grant application, but consultants normally charge money for their services. Federal grant programs are published in

TEXT BOX 13-2

COMMON EMERGENCY SERVICE–RELATED GRANT PROGRAMS

- *Assistance to Firefighters Grant Program (AFGP).* Administered by the Department of Homeland Security's Office for Domestic Preparedness, this program assists rural, urban, and suburban fire departments throughout the United States. These funds are used by the nation's firefighters to increase the effectiveness of firefighting operations, to improve firefighter health and safety programs, and to establish or expand fire prevention and safety programs. This program was known as the Fire Investment and Response Enhancement (FIRE) Act and was moved from FEMA to the Department of Homeland Security in 2004. This grant program also includes funding for fire prevention programs and has a grant program for staffing. The staffing grant program is the Staffing for Adequate Fire & Emergency Response Grants (SAFER) program.

- *Department of Justice.* Has grants available for terrorism-preparedness equipment and training.

- *Volunteer Fire Assistance (VFA).* The purpose of this program, formerly known as the Rural Community Fire Protection (RCFP) Program, is to provide federal financial, technical, and other assistance to state foresters and other appropriate officials to organize, train, and equip fire departments in rural areas and rural communities to prevent and suppress fires. A rural community is defined as having a population of 10,000 or less.

- *Hazardous Materials Emergency Preparedness (HMEP).* This grant program is intended to provide financial and technical assistance as well as national direction and guidance to enhance state, territorial, tribal, and local hazardous materials emergency planning and training. The HMEP grant program distributes fees collected from shippers and carriers of hazardous materials to emergency responders for hazmat training and to local emergency planning committees (LEPCs) for hazmat planning.

the *Federal Register.* Safety and health managers should focus on the proposed project and look at both emergency service–based grants as well as those outside of the emergency field. A department received a technology grant for laptop computers and a GIS mapping and preplanning system for its apparatus. This enhanced information was available at the emergency scene, hence it helped enhance safety.

Often the grant applicant must show how funds will be allocated to continue the program after the grant period ends. Some grants require a local match, some do not. Some examples of emergency services–related grant programs are listed in **Text Box 13-2**. This is a brief example list only and should not be considered all inclusive.

SAFETY CONSIDERATIONS IN THE DIVERSIFIED WORKFORCE

According to the report *Workforce 2000*,[1] diversity in the workplace will continue to increase. The emergency services have seen these changes and will continue to. Although this document was written for the workforce as a whole, there are implications for the emergency service occupations. Some of the findings from the report include:

- Population growth will slow.
- The average age of the workforce will rise.
- More women will enter the workforce.
- White males will be a smaller percentage of the workforce.
- Minorities will be a larger share of the new entrants into the labor force.

Although there are not safety and health considerations for all of these findings, some of these findings warrant further discussion.

NOTE The increasing average age of the workforce will require continued emphasis on physical fitness.

Physical abilities testing should be based on job-validated requirements. The IAFC/IAFF joint fitness and wellness initiative group has developed one such abilities test, the Candidate Physical Abilities Test, or CPAT. The CPAT validated abilities test consists of a timed course evaluating performance on eight events. Information on the CPAT is available from the IAFC or through IAFF locals.

Although the emergency services have generally adapted well to women in the workforce, their entry has been relatively recent. What safety and health considerations must be undertaken for women who are pregnant? Policies must be developed before the situation arises. Many chemicals can affect the unborn fetus as well as the mother, and heavy lifting may also be a concern.

The design and location of equipment on apparatus must be considered. Part of a diverse workforce is variations in height. Years ago, fire departments had height requirements. After these were challenged and removed, there became a need to redesign apparatus, particularly ladder storage, patient compartment height, and the height of hose beds.

Equipment purchase recommendations should consider the diversity in personnel. OSHA requires fit testing of SCBA masks. It is unlikely that one size of masks on the apparatus would provide a proper fit for all members. The same holds true for other PPE; the organization must make available a variety of sizes of PPE, as protective clothing must fit well and comfortably without causing undue heat stress or adding excessive weight.

NOTE A firefighter with improperly fitting protective equipment will be inefficient and may be more prone to injury.

These are but a few examples of safety and health considerations in a diversified workforce. The safety and health manager must maintain awareness of trends in these areas and be able to adjust the safety and health program as needed.

FUTURE TRENDS AND TECHNOLOGIES

Probably one of the more exciting prospects in the field of safety and health is looking toward the future. Just consider for a moment the technological changes that have occurred over the past 20 years. Look at the protective equipment changes and how the emergency services have had to adapt. Many items are being manufactured now that are somewhat expensive and out of reach of some departments but, as technology improves, the market will drive the price down. Remember the first personal computers? Compare their price and capabilities to what you can buy today for a fraction of the cost and size. Ten years ago, few departments had a thermal imaging camera. Now that costs have come down some and additional grant funding has become available, many departments have one camera on each apparatus. Some issues that could impact the safety and health program in the future are given in the following list:

- Emergency Scene Geographic Information System (GIS) interface with the GPS linked to the incident commander's computer for location of all personnel.
- Recruitment and retention of qualified personnel.
- Terrorism preparedness causing a greater need to work closely with other related response agencies and creating the need for interoperability of communications equipment and better coordination of IMS.
- Individual status information projected on the SCBA mask and wirelessly to command officers.
- The ability to have building floor plans projected on the SCBA facemask.
- Better early warning for fires and better private fire protection.
- Greater use of GPS and intersection control to reduce intersection accidents.
- Even greater use of thermal imaging systems.
- Further improvements in protective clothing to prevent exposures.
- Greater emphasis on occupational safety. Hiring of specific health and safety officers with higher-level degrees in occupational safety and health.
- A continued national interest in emergency service safety and health.
- More emergency service safety and health-related education and training.
- A greater emphasis on research for health and safety programs including a study of what is done in other countries, giving the safety and health problem a global aspect.

SUMMARY

The safety and health concepts presented in this book would not be complete without a look at legal, ethical, and financial issues. The legal issues are best dealt with using legal counsel; however, the safety and health program manager should have a working knowledge of some legal concepts, specifically those involving negligence and compliance with applicable federal regulations. Ethics are a part of every organization. Members active with the safety and health program may be held to a higher ethical standard as a result of their position. The financial issues involve funding for recurring annual costs as well as costs associated with program component startup. The safety and health manager should have input into the budget process. The annual budget is the most common funding source, but alternative funding sources, such as grants, do exist.

Changes in the diversity and demographics will have implications to the safety and health program manager and to the overall program. These changes have been identified in the *Workforce 2000* document and focus on a slowing population growth, a rise in the average age on the workforce, more women entering the workforce, white males becoming a smaller percentage of the workforce, and minorities becoming a larger share of the new entrants into the workforce. There are health and safety considerations in a couple of these areas.

Future technology and trends will affect the safety and health program. From better protective equipment to the safety and health manager becoming a highly trained professional similar to a fire protection engineer, the focus will be on continued support and improvement of the safety and health for the emergency responders of the world.

Concluding Thought: There are a number of special considerations in the development and management of the safety and health program. The safety and health officer must have a working knowledge in a variety of subjects broader than just emergency service operations.

REVIEW QUESTIONS

1. List the predictions of *Workforce 2000* and identify those that will have a health and safety impact.
2. Describe two funding sources for the safety and health program.
3. List five implications of future technology of a safety and health program.
4. Describe the four kinds of tort liability presented in this textbook.
5. What are the implications of an older workforce on the health and safety program?

ACTIVITIES

1. Examine the *Workforce 2000* predictions. How have they affected your department?
2. Review any grant funds that your department has received. If none, what opportunities do you see?
3. Pick any two future technologies or trends and discuss their implications for your department.

NOTES

1. Published in 1987 by the Hudson Institute of Indianapolis for the Department of Labor.

Reprint of the 2010 Firefighter Fatality Survey from the United States Fire Administration

Although this text only includes the report from 2010, these reports are done annually. The report is included here to introduce the reader to the information available from a report of this nature. The reader is encouraged to obtain a current report from the publications center of the USFA or from http://www.usfa.fema.gov for local comparison. The complete report includes an appendix that is a summary of individual incidents. This summary is not included here. (Courtesy of Federal Emergency Management Agency (FEMA))

U.S. Fire Administration

Firefighter Fatalities in the United States in 2010

September 2011

 FEMA

U.S. Fire Administration
Mission Statement

We provide National leadership to foster a solid foundation for local fire and emergency services for prevention, preparedness and response.

Firefighter Fatalities in the United States in 2010

Prepared by

U.S. Department of Homeland Security

Federal Emergency Management Agency

U.S. Fire Administration

National Fire Data Center

and

The National Fallen Firefighters Foundation

www.firehero.org

In memory of all firefighters
who answered their last call in 2010
To their families and friends
To their service and sacrifice

"We will never forget"

Cover photo courtesy of Rae Stewart.

TABLE OF CONTENTS

ACKNOWLEDGMENTS

This study of firefighter fatalities would not have been possible without the cooperation and assistance of many members of the fire service across the United States. Members of individual fire departments, chief fire officers, wildland fire service organizations such as the U.S. Forest Service (USFS), the National Park Service (NPS), the Bureau of Land Management (BLM), the Bureau of Indian Affairs (BIA), the U.S. Fish and Wildlife Service (FWS), as well as the U.S. Department of Justice (DOJ), the National Fire Protection Association (NFPA), and many others contributed important information to this report.

The National Fallen Firefighters Foundation (NFFF) was responsible for compilation of a large portion of the data used in this report and the incident narrative summaries found in Appendix A.

The ultimate objective of this effort is to reduce the number of firefighter deaths through an increased awareness and understanding of their causes and how they can be prevented. Firefighting, rescue, and other types of emergency operations are essential activities in an inherently dangerous profession, and unfortunate tragedies do occur. These are the risks all firefighters accept every time they respond to an emergency incident. However, the risks can be greatly reduced through efforts to improve training, emergency scene operations, and firefighter health and safety initiatives.

BACKGROUND

For 34 years, the U.S. Fire Administration (USFA) has tracked the number of firefighter fatalities and conducted an annual analysis. Through the collection of information on the causes of firefighter deaths, the USFA is able to focus on specific problems and direct efforts toward finding solutions to reduce the number of firefighter fatalities in the future. This information is also used to measure the effectiveness of current programs directed toward firefighter health and safety.

Several programs have been funded by USFA in response to this annual report. For example, USFA has sponsored significant work in the areas of general emergency vehicle operations safety, fire department tanker/tender operations safety, firefighter incident scene rehabilitation, and roadside incident safety. The data developed for this report are also widely used in other firefighter fatality prevention efforts.

One of USFA's main program goals is a 25-percent reduction in firefighter fatalities in 5 years and a 50-percent reduction within 10 years. The emphasis placed on these goals by USFA is underscored by the fact that these goals represent one of the five major objectives that guide the actions of USFA.

In addition to the analysis, USFA, working in partnership with the NFFF, develops a list of all onduty firefighter fatalities and associated documentation each year. If certain criteria are met, the fallen firefighter's next of kin, as well as members of the individual's fire department, are invited to the annual National Fallen Firefighters Memorial Weekend Service. The service is held at the National Emergency Training Center (NETC) in Emmitsburg, MD, during Fire Prevention Week in October of each year. Additional information regarding the Memorial Service can be found at www.firehero.org or by calling the NFFF at (301) 447-1365.

Other resources and information regarding firefighter fatalities, including current fatality notices, the National Fallen Firefighters Memorial database, and links to the Public Safety Officers' Benefit (PSOB) Program can be found at www.usfa.dhs.gov/fireservice/fatalities/

INTRODUCTION

This report continues a series of annual studies by the U.S. Fire Administration (USFA) of onduty firefighter fatalities in the United States.

The specific objective of this study is to identify all onduty firefighter fatalities that occurred in the United States and its protectorates in 2010 and to analyze the circumstances surrounding each occurrence. The study is intended to help identify approaches that could reduce the number of firefighter deaths in future years.

Who is a Firefighter?

For the purpose of this study, the term firefighter covers all members of organized fire departments with assigned fire suppression duties in all 50 States, the District of Columbia, and the Territories of Puerto Rico, the Virgin Islands, American Samoa, the Commonwealth of the Northern Mariana Islands, and Guam. It includes career and volunteer firefighters; full-time public safety officers acting as firefighters; fire police; State, territory, and Federal government fire service personnel, including wildland firefighters; and privately employed firefighters, including employees of contract fire departments and trained members of industrial fire brigades, whether full or part time. It also includes contract personnel working as firefighters or assigned to work in direct support of fire service organizations (i.e., air-tanker crews).

Under this definition, the study includes not only local and municipal firefighters, but also seasonal and full-time employees of the U.S. Forest Service (USFS), the National Park Service (NPS), the Bureau of Land Management (BLM), the Bureau of Indian Affairs (BIA), the U.S. Fish and Wildlife Service (FWS), and State wildland agencies. The definition also includes prison inmates serving on firefighting crews; firefighters employed by other governmental agencies, such as the U.S. Department of Energy (DOE); military personnel performing assigned fire suppression activities; and civilian firefighters working at military installations.

What Constitutes an Onduty Fatality?

Onduty fatalities include any injury or illness sustained while on duty that proves fatal. The term "on duty"

refers to being involved in operations at the scene of an emergency, whether it is a fire or nonfire incident; responding to or returning from an incident; performing other officially assigned duties such as training, maintenance, public education, inspection, investigations, court testimony, and fundraising; and being on call, under orders, or on standby duty except at the individual's home or place of business. An individual who experiences a heart attack or other fatal injury at home while he or she prepares to respond to an emergency is considered on duty when the response begins. A firefighter that becomes ill while performing fire department duties and suffers a heart attack shortly after arriving home or at another location may be considered on duty since the inception of the heart attack occurred while the firefighter was on duty.

On December 15, 2003, the President of the United States signed into law the Hometown Heroes Survivors Benefit Act of 2003. After being signed by the President, the Act became Public Law 108-182. The law presumes that a heart attack or stroke are in the line of duty if the firefighter was engaged in nonroutine stressful or strenuous physical activity while on duty and the firefighter becomes ill while on duty or within 24 hours after engaging in such activity. The full text of the law is available at http://frwebgate.access.gpo.gov/cgi-bin/getdoc.cgi?dbname=108_cong_public_laws&docid=f:publ182.108.pdf

The inclusion criteria for this study have been affected by this change in the law. Previous to December 15, 2003, firefighters who became ill as the result of a heart attack or stroke after going off duty needed to register a complaint of not feeling well while still on duty in order to be included in this study. For firefighter fatalities after December 15, 2003, firefighters will be included in this report if they became ill as the result of a heart attack or stroke within 24 hours of a training activity or emergency response. Firefighters who became ill after going off duty where the activities while on duty were limited to tasks that did not involve physical or mental stress will not be included.

A fatality may be caused directly by an accidental or intentional injury in either emergency or nonemergency circumstances, or it may be attributed to an occupationally related fatal illness. A common example of a fatal illness incurred on duty is a heart attack. Fatalities attributed to occupational illnesses also include a communicable disease contracted while on duty that proved fatal when the disease could be attributed to a documented occupational exposure.

Firefighter fatalities are included in this report even when death is considerably delayed after the original incident. When the incident and the death occur in different years, the analysis counts the fatality as having occurred in the year in which the incident took place.

There is no established mechanism for identifying fatalities that result from illnesses such as cancer that develop over long periods of time and which may be related to occupational exposure to hazardous materials or toxic products of combustion. It has proved to be very difficult to provide a complete evaluation of an occupational illness as a causal factor in firefighter deaths due to the following limitations: the exposure of firefighters to toxic hazards is not sufficiently tracked; the often delayed long-term effects of such toxic hazard exposures; and the exposures firefighters may receive while off duty.

Sources of Initial Notification

As an integral part of its ongoing program to collect and analyze fire data, the USFA solicits information on firefighter fatalities directly from the fire service and from a wide range of other sources. These sources include the Public Safety Officers' Benefit (PSOB) Program administered by the U.S. Department of Justice (DOJ), the National Institute for Occupational Safety and Health (NIOSH), the Occupational Safety and Health Administration (OSHA), the Department of Defense (DOD), the National Interagency Fire Center (NIFC), and other Federal agencies.

The USFA receives notification of some deaths directly from fire departments, as well as from such fire service organizations as the International Association of Fire Chiefs (IAFC), the International Association of Fire Fighters (IAFF), the National Fire Protection Association (NFPA), the National Volunteer Fire Council (NVFC),

State fire marshals, State fire training organizations, other State and local organizations, fire service Internet sites, news services, and fire service publications.

Procedure for Including a Fatality in the Study

In most cases, after notification of a fatal incident, initial telephone contact is made with local authorities by the USFA to verify the incident, its location, jurisdiction, and the fire department or agency involved. Further information about the deceased firefighter and the incident may be obtained from the chief of the fire department, designee over the phone, or by other forms of data collection. After basic information is collected, a notice of the firefighter fatality is posted at the National Fallen Firefighters Memorial site in Emmitsburg, MD, the USFA website, and a notice of the fatality is transmitted by electronic mail to a large list of fire service organizations and fire service members.

Information that is routinely requested from fire departments that have experienced a fatality includes National Fire Incident Reporting System (NFIRS)-1 (incident) and NFIRS-3 (fire service casualty) reports; the fire department's own incident and internal investigation reports; copies of death certificates and autopsy results; special investigative reports; law enforcement reports; photographs and diagrams; and newspaper or media accounts of the incident. Information on the incident may also be gathered from NFPA or NIOSH reports.

After obtaining this information, a determination is made as to whether the death qualifies as an on-duty firefighter fatality according to the previously described criteria. With the exception of firefighter deaths after December 15, 2003, the same criteria were used for this study as in previous annual studies. Additional information may be requested by USFA, either through follow-up with the fire department directly, from State vital records offices, or other agencies. The final determination as to whether a fatality qualifies as an onduty death for inclusion in this statistical analysis is made by the USFA and the Foundation's criteria as a line-of-duty death (LODD) for inclusion in the annual National Fallen Firefighters Memorial Service is made by the National Fallen Firefighters Foundation (NFFF).

2010 FINDINGS

Eighty-seven firefighters died while on duty in 2010, continuing a second year of substantially fewer firefighter deaths in the United States, where, during the previous 6-year period of 2004–2009, the average number of annual onduty firefighter deaths was 112.

The 2010 total includes 15 firefighters who died under circumstances as a result of inclusion criteria changes resulting from the Hometown Heroes Act of 2003. When not including these fatalities in a trend analysis, the 2010 total of 72 firefighter fatalities was the lowest number of firefighter losses on record over the past 34 years.

An analysis of multiyear firefighter fatality trends needs to acknowledge the changes from the Home-

town Heroes Survivors Benefit Act of 2003; therefore, some graphs and charts either will or will not indicate the Hometown Heroes portion of the total. However, this does not diminish the sacrifices made by any firefighter who dies while on duty or the sacrifices made by his/her family and peers.

Moreover, when conducting multiyear comparisons of firefighter fatalities in this report, the losses that were the result of the attacks on the World Trade Center (WTC) in New York City on September 11, 2001, are sometimes also set apart for illustrative purposes. This action is by no means a minimization of the supreme sacrifice made by these firefighters.

Figure 1. Onduty Firefighter Fatalities (1977–2010).

Figure 2. Firefighter Fatalities per 100,000 Fires.

*2010 ratio will be included in the 2011 report.

Career, Volunteer, and Wildland Agency Deaths

In 2010, firefighter fatalities included 56 volunteer firefighters, 28 career firefighters, and three part-time or full-time members of wildland or wildland contract fire agencies (Figure 3).

Figure 3. Career, Volunteer, and Wildland Agency Deaths (2010).

Firefighter Fatalities
87 Total Deaths

Career
28 Deaths

Volunteer
56 Deaths

Wildland Agency
3 Deaths

Rural Fire
Departments
5 Death

Rural Fire
Departments
39 Deaths

Full time
2 Deaths

Urban/Suburban
Fire Departments
23 Deaths

Urban/Suburban
Fire Departments
17 Deaths

Part time or
Seasonal
1 Death

Gender

All of the firefighters who died while on duty in 2010 were male. This is the first year since 1998 where no female firefighters were killed while on duty. During the previous decade, 2000–2009, there were an average of four female, and not including WTC, 106 male onduty firefighter deaths per year.

Multiple Firefighter Fatality Incidents

The 87 deaths in 2010 resulted from a total of 83 fatal incidents. There were four firefighter fatality incidents where two firefighters were killed in each in 2010, claiming a total of eight firefighters.

Table 1. Multiple Firefighter Fatality Incidents

Year	Number of Incidents	Total Number of Deaths
2010	4	8
2009	6	13
2008	5	18
2007	7	21
2006	6	17
2005	4	10
2004	3	6
2003	7	20
2002	9	25
2001	8	362
2001 w/o WTC	7	18

Wildland Firefighting Deaths

In 2010, 11 firefighters were killed during activities involving brush, grass, or wildland firefighting. This total includes part-time and seasonal wildland firefighters, full-time wildland firefighters, and municipal or volunteer firefighters whose deaths are related to a wildland fire (Figure 4).

Figure 4. Firefighter Fatalities Related to Wildland Firefighting (2001–2010).

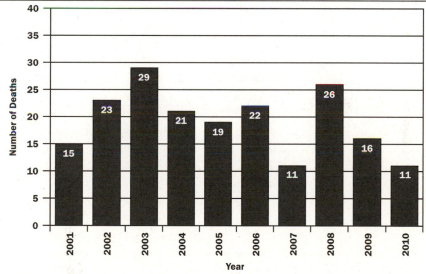

Table 2. Wildland Firefighting Aircraft Deaths

Year	Total Number of Deaths	Number of Fatal Incidents
2010	0	0
2009	5	3
2008	16	4
2007	1	1
2006	8	3
2005	6	2
2004	3	3
2003	7	4
2002	6	3
2001	6	3

In 2010, there were no multiple firefighter fatality incidents related to wildland firefighting and, for the first time since 1999, no wildland firefighter deaths related to aircraft.

Table 3. Firefighter Deaths Associated with Wildland Firefighting

Year	Total Number of Deaths	Number of Fatal Incidents	Number of Firefighters Killed in Multiple-Death Incidents
2010	11	11	0
2009	16	13	5
2008	26	15	14
2007	11	11	0
2006	22	13	13
2005	19	15	6
2004	21	21	0
2003	30	22	10
2002	23	14	13
2001	15	9	9

TYPE OF DUTY

Activities related to emergency incidents resulted in the deaths of 48 firefighters in 2010 (Figure 5). This includes all firefighters who died responding to an emergency or at an emergency scene, returning from an emergency incident, and during other emergency-related activities. Nonemergency activities accounted for 39 fatalities. Nonemergency duties include training, administrative activities, performing other functions that are not related to an emergency incident, and postincident fatalities where the firefighter does not experience the illness or injury during the emergency.

Figure 5. Firefighter Deaths by Type of Duty (2010).

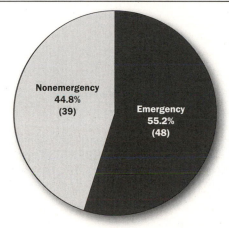

A multiyear historical perspective relating to the percentage of firefighter deaths that occurred during emergency duty is presented in Table 4.

Table 4. Emergency Duty Firefighter Deaths

Year	Percentage of All Deaths	Percentage of All Deaths Without Hometown Heroes
2010	55.2	66.7
2009	63.3	82.2
2008	63.5	70.0
2007	64.4	72.4
2006	57.5	66.3
2005	52.1	60.6
2004	68.9	75.9
2003	69.0	69.6
2002	73.0	N/A
2001	65.0	N/A
2001 with WTC	92.0	N/A
2000	71.0	N/A
1999	87.0	N/A
1998	77.0	N/A

The number of deaths by type of duty being performed in 2010 is shown in Table 5 and presented graphically in Figure 6. As has been the case for most years, fireground duties are the most common type of duty for firefighters killed while on duty.

Table 5. Firefighter Deaths by Type of Duty (2010)

Type of Duty	Number of Deaths
Fireground Operations	22
Other Onduty Deaths	18
Responding	15
Returning	1
Training	12
Onscene Nonfire Emergencies	4
After	15
Total	87

Figure 6. Firefighter Deaths by Type of Duty (2010).

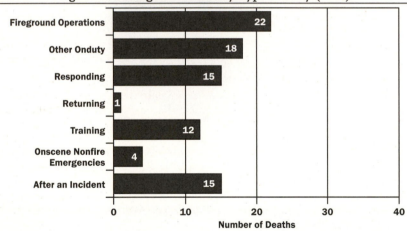

Fireground Operations

Of the 22 firefighters killed during fireground operations in 2010, 14 were at the scene of a structure fire, and 8 others were at the scene of a wildland or outside fire. The average age of the firefighters killed during fireground operations was 45, half of whom were from career fire departments, just under half (10) from volunteer fire departments, and one firefighter from a wildland agency.

Type of Fireground Activity

Table 6 shows the types of fireground activities in which firefighters were engaged at the time they sustained their fatal injuries or illnesses. This total includes all firefighting duties, such as wildland firefighting and structural firefighting.

Table 6. Type of Activity (2010)

Water Supply	1
Pump Ops	1
Search & Rescue	3
Advance Hoselines	10
Other	2
Unknown	4
Incident Command	1

Fixed Property Use for Structural Firefighting Deaths

There were 14 fatalities in 2010 where firefighters became ill or injured while on the scene of a structure fire. Table 7 shows the distribution of these deaths by fixed property use.

Table 7. Structural Firefighting Deaths by Fixed Property Use in 2010

Residential	8
Commercial	6

Responding/Returning

Sixteen firefighters died while responding to or returning from 15 emergency incidents in 2010. Fifteen of the firefighters died while responding to incidents and one while returning from an incident.

Nine of the firefighters killed while responding to incidents died from heart attacks (eight) or a stroke (one).

Six of the firefighters who died while responding to incidents were killed by trauma caused by motor vehicle collisions, including three in privately-owned vehicles (POVs) and three in fire department apparatus. Of the POV crashes taking a firefighter's life, one occurred on an all-terrain vehicle (ATV) that struck a deer. No helmet and alcohol, the only 2010 incident where alcohol was reported to be involved, were contributing factors in the death. In the two other separate POV incidents, seat restraints were present in both vehicles. One firefighter was wearing them and one was not; the latter fully ejected from the vehicle.

Of the three firefighters killed in fire department apparatus involved in motor vehicle collisions, two were not wearing seat restraints and fully ejected when their engine was struck at an intersection by a sport utility vehicle (SUV). The third firefighter killed, who was belted in and not ejected from the vehicle, was operating a semitrailer truck hauling a D5 Caterpillar plow that did not successfully negotiate a curve. As a result of the accident, the firefighter became trapped in the cab of the truck and was pronounced dead at the scene.

The one firefighter killed while returning from an incident was crushed between a moving tanker and a parked apparatus as the tanker backed into the station's apparatus bay.

Table 8. Firefighter Deaths While Responding to or Returning from an Incident

Year	Number of Firefighter Deaths
2010	16
2009	15
2008	24
2007	26
2006	15
2005	22
2004	23
2003	36
2002	13
2001	23

Training

In 2010, 12 firefighters died while engaged in training activities. Seven of the deaths were due to heart attacks, two from cerebrovascular accidents (CVA), and one from a documented case of necrotizing fasciitis (commonly known as flesh eating disease). An additional two firefighters were killed in a POV while they were returning from a fire certification class at a local community college. These firefighters were apparently involved in street racing with firefighters in another vehicle.

Table 9. Firefighter Fatalities While Engaged in Training

Year	Number of Firefighter Deaths
2010	12
2009	10
2008	12
2007	11
2006	9
2005	14
2004	13
2003	12
2002	11
2001	14

Nonfire Emergencies

In 2010, there were two fire police and two firefighter fatalities where the type of emergency duty was not related to a fire. One fire police was struck and killed by a vehicle while performing scene safety duties at the scene of a motor vehicle accident. Another fire police suffered a heart attack after establishing a medivac helispot to evacuate a patient from an emergency medical services (EMS) incident. The two firefighters both died from asphyxiation in separate incidents. The first firefighter was part of a three-person crew in a boat and actively engaged in rescue operations when the boat was captured by strong currents, struck a bridge, and capsized. The second firefighter died as a result of oxygen deprivation in a sewer system manhole while he attempted to rescue a member of the town's Department of Public Works (DPW) who had lost consciousness after descending the interior ladder and falling to the bottom.

After the Incident

In 2010, 15 firefighters died after the conclusion of their onduty activity. All 15 deaths were due to heart attacks.

CAUSE OF FATAL INJURY

The term "cause of injury" refers to the action, lack of action, or circumstances that directly resulted in the fatal injury. The term "nature of injury" refers to the medical cause of the fatal injury or illness which is often referred to as the physiological cause of death. A fatal injury is usually the result of a chain of events, the first of which is recorded as the cause.

In 2010, one firefighter was reported to have committed suicide while on duty.

Figure 7 shows the distribution of deaths by cause of fatal injury or illness in 2010.

Figure 7. Fatalities by Cause of Fatal Injury (2010).

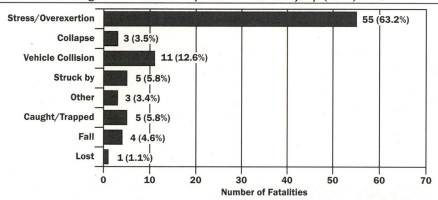

Stress or Overexertion

Firefighting is extremely strenuous physical work and is likely one of the most physically demanding activities that the human body performs.

Stress or overexertion is a general category that includes all firefighter deaths that are cardiac or cerebrovascular in nature such as heart attacks, strokes, and other events such as extreme climatic thermal expo-

sure. Classification of a firefighter fatality in this cause of fatal injury category does not necessarily indicate that a firefighter was in poor physical condition.

Fifty-five firefighters died in 2010 as a result of stress/overexertion:

- Fifty firefighters died due to heart attacks.

- Five firefighters died due to CVAs.

Table 10. Deaths Caused by Stress or Overexertion

Year	Number	Percent of Fatalities
2010	55	63.2
2009	50	55.5
2008	52	44.0
2007	55	46.6
2006	54	50.9
2005	62	53.9
2004	66	56.4
2003	51	45.9**
2002	38	38.0
2001	43	40.9*

*Does not include the firefighter deaths of September 11, 2001, in New York City.
**Includes Hometown Heroes, one in December 2003, and an average of 13.4 for the years 2004–2010.

Vehicle Crashes

Eleven firefighters died in 2010 as the result of vehicle crashes, down 31 percent from vehicle crashes in 2009 and almost 61 percent from 2008. In 2010, there were no firefighter deaths that involved aircraft.

In addition to the responding/returning from an incident and training-related vehicle crash fatalities earlier in this report, there were two additional vehicle crashes that killed firefighters in 2010:

- While on scene of a mutual-aid wildland fire call, a nontraffic vehicle (snowcat) ascended a hill, lost traction, and began to tumble down the slope, killing the operator as it rolled more than three times. Another firefighter was ejected from the snowcat but survived the accident.

- A wildland crew carrier, with as many as 12 firefighters in the vehicle, was struck by an oncoming SUV that had crossed over the center line of the highway. Several firefighters were ejected in the crash and a number received severe injuries, including one who died from his injuries.

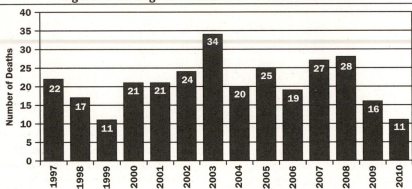

Figure 8. Firefighter Fatalities in Vehicle Collisions.

Lost or Disoriented

One firefighter died in 2010 when he became lost or disoriented inside of a burning residential structure while searching for an elderly couple and after having rescued their dog, was reported to still be inside of a burning residential structure. Rapid Intervention Team (RIT) crews were assigned and searched the structure for the firefighter. After approximately 10 minutes of searching by multiple crews, he was located in a small room behind a closed door. Investigation revealed that the firefighter had become ill and vomited into his self-contained breathing apparatus (SCBA) facepiece. When he was found, he was lying on his back without his helmet, gloves, and facepiece. His death was caused by smoke inhalation.

Caught or Trapped

Five firefighters were killed in 2010 in four separate incidents when they became caught or trapped. This classification covers firefighters trapped in wildland and structural fires who were unable to escape due to rapid fire progression and the byproducts of smoke, heat, toxic gases, and flame. This classification also includes firefighters who drowned and those who were trapped and crushed.

- Firefighters conducting search activities in a residential structure observed fire advancement and yelled to the hoseline crew to evacuate. Once firefighters exited, they found that two firefighters were still in the structure. Fire conditions inside had dramatically changed. One firefighter was able to make it to within a few feet of the door as he was pulled from the structure by other firefighters. Firefighters entered the structure with a hoseline to search for the other missing firefighter. He was found wrapped up in the ruptured 2-1/2 inch handline and, although wearing his facepiece when he entered the structure, was not wearing his facepiece, hood, or helmet. The firefighter was removed from the structure, treated by other firefighters, and transported to the hospital by ambulance, but did not survive.

- One firefighter and his engine company arrived approximately 20 minutes after dispatch and found a working fire in a large egg production and processing facility. Two firefighters advanced an attack line into the structure but became separated from the hoseline and then from each other in the interior. One firefighter was able to find a metal exterior wall and banged on it until firefighters on the exterior were able to cut open the wall to allow his escape. The fire progressed and attempts to locate the second firefighter were unsuccessful. He died of thermal injuries and smoke inhalation.

- Upon their arrival at the scene, firefighters found an active fire on the second floor of a residential structure. Two firefighters were assigned to go to the third floor of the building to look for fire extension and to search for victims. When both firefighters, who were wearing full structural protective clothing, including SCBAs, arrived on the third floor, one firefighter reported stable conditions by radio to the Incident Commander (IC) and requested that a hoseline be brought to the floor. A few moments later, the IC observed smoke coming from the third floor and attempted to contact the firefighter by radio to no avail. The IC ordered an evacuation of the building by all firefighters in order to account for everyone on the scene. In the process of the evacuation, firefighters were observed on the third floor in distress. The RIT was deployed to the third floor for rescue. Both firefighters were found unconscious by firefighters and brought to the exterior of the building. Firefighters and emergency responders immediately began to provide medical treatment for both firefighters and they were transported to the hospital but did not survive their injuries. Autopsies and toxicological studies showed that both died of smoke inhalation suffered as they fought the structure fire.

- Firefighters were requested to assist Department of Public Works (DPW) employees with access to a sewer system manhole behind their fire station. A DPW employee entered the manhole and lost consciousness as he descended the interior ladder and fell to the bottom. The DPW foreperson requested assistance from firefighters. The fire chief asked dispatch for EMS assistance and also started a fire

department incident. As an atmospheric meter from the fire department was prepared for use, a firefighter entered the opening. He was not wearing a SCBA or harness. When the firefighter was halfway down the ladder, he lost consciousness and fell to the bottom. Additional resources were called to the scene and both the firefighter and the DPW employee were removed from the opening; both died of asphyxiation due to an oxygen deficient atmosphere.

Collapse

Three firefighters died in 2010 as a result of structural collapses in two separate commercial property fire incidents.

- Firefighters found a working fire in an office building and initiated an interior attack. Firefighters found fire in the attic of the structure and began to evacuate the interior. Approximately 30 minutes into the fire fight, two firefighters were operating a handline on the exterior of the structure when a collapse occurred. One firefighter was trapped in debris and was removed by other firefighters using hydraulic tools. He was transported to a local hospital and was treated for crush and thermal injuries. He was transported to a regional hospital and then to a care facility where he died almost 5 months later from injuries sustained.

- A small, intentionally-set rubbish fire had been quickly extinguished with tank water and extension into the trusses was ruled out in a vacant commercial building. The east wall (sector 4 side) of the structure had three large wire glass metal framed industrial-type windows. Firefighters began removing these windows to provide a quick means of egress for firefighters working within the structure. As the southern-most window on the east wall was removed, it is theorized that the entire east wall failed. The east wall was a bearing wall which supported the weight of two heavy timber bowstring trusses. The failure of the east wall caused total collapse of the bowstring truss roof and pushed out the sector 3 wall. Sixteen firefighters became trapped and injured in the collapse. Two firefighters passed away from injuries sustained in this incident.

Struck by Object

Being struck by an object was the cause of five fatal firefighter injuries in 2010.

- One fire police was working scene safety and traffic control services for a mutual-aid motor vehicle crash with entrapment and wires down. He parked his vehicle on the side of the road with its warning lights in operation and set cones and a flare in the roadway to warn approaching drivers. A civilian vehicle disregarded the road closure, drove over cones, and struck the fire police officer. He was transported to the hospital but was pronounced dead upon arrival.

- One firefighter was clearing a fire break for containment at the base of a bluff when a burning snag broke loose on top and rolled downhill over a small bluff, striking him from behind. The firefighter sustained a serious head injury, fractured hip, bruises, and second-degree burns on his calves. The impact left him unconscious and with serious injuries, including the burns from which he did not recover.

- One firefighter was staffing a brush truck while fighting a wildland fire when the water tank on the brush truck failed under pressure and the firefighter was struck in the chest by a piece of the tank. The firefighter was killed as a result of trauma and was pronounced dead at the scene.

- A firefighter/mechanic was making steering system repairs to a brush truck inside of the fire department's maintenance building. As the mechanic turned the key to unlock the steering column, the brush truck accidentally lunged forward. At the same time the brush truck lunged forward, two firefighters were walking in front of the vehicle. One of the firefighters was knocked clear but the other was crushed between the brush truck and a wall. He received a head injury and died instantly from his injuries.

- Two engine companies were dispatched to a fire in the median of a local highway. Upon arrival, engine personnel positioned their apparatus in a shielding position to protect firefighters from traffic. As firefighters were extinguishing the fire, a vehicle crash occurred on the highway behind the apparatus. The crash caused one of the vehicles to slide between the guardrail and one of the engines. The vehicle entered the fire scene and struck two firefighters. Both firefighters were transported to the hospital but one of them died due to traumatic injuries.

Fall

Four firefighters died in 2010 as the result of falls.

- After a firefighter had closed and latched the doors at the top of a burning grain silo, he fell 20–30 feet when an explosion occurred inside of the silo. The firefighter was transported to the hospital by ambulance but was pronounced dead at the hospital due to traumatic injuries.

- A firefighter working in-station duties experienced a fall. Firefighters treated him for a cut on his head and recommended that he go to the hospital to be checked. Later that night, they checked on their fellow firefighter in the hospital and found him in better condition than before. The firefighter, however, later took a turn for the worse and died from the head injury received in the fall at the fire station.

- A firefighter and his ladder company were dispatched to a structural fire in a four-story commercial and residential structure. Upon their arrival, firefighters observed smoke and sparks coming from a cooking exhaust fan chute. The firefighter ascended a ladder attached to the side of the building to gain access to the roof. He was wearing full structural firefighting protective clothing, including SCBA. He also carried a water fire extinguisher with him as he climbed. When the firefighter reached the roof, he lost his grip and fell 53 feet to the ground. The firefighter was severely injured in the fall and was treated by firefighters and transported to the hospital. He was pronounced dead at the hospital due to multiple injuries.

- Members of a fire department were working on the construction of a new fire station to replace one that had been destroyed by fire. One firefighter died from injuries received on the worksite when he fell backwards and struck his head on concrete.

Other

Three firefighters died in 2010 of a cause that is not categorized above.

- One firefighter was struck by an Amtrak train. State police, along with Amtrak police and the city, investigated the circumstances surrounding the death. No foul play was suspected.

- One firefighter was a part of a three-person crew in a boat and actively engaged in rescue operations when it was captured by strong currents, struck a bridge, and capsized.

- During the course of a training exercise, a firefighter struck his elbow on a portion of the body of an engine as he removed SCBA cylinders for use in the training. He noted his injury to other firefighters but thought that it was insignificant. The firefighter signed out of training but approximately 6 hours later awoke and began vomiting. He was also experiencing pain is his injured elbow. An ambulance was called. Arriving EMS responders treated the incident as a cardiac event based on symptoms and transported him to the hospital. The firefighter was diagnosed as suffering from necrotizing fasciitis (commonly known as flesh eating disease). He underwent multiple surgeries to attempt to control the spread of the disease and was transferred to a burn unit for treatment. The firefighter underwent a total of 18 operations during 79 days in the hospital. Although his condition seemed to improve for a time, he died as a result of complications related to the injury.

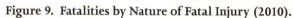

NATURE OF FATAL INJURY

Figure 9 shows the distribution of the 87 firefighter deaths that occurred in 2010 by the medical nature of the fatal injury or illness. For heart attacks, Figure 10 shows the type of duty involved.

> The National Institute for Occupational Safety and Health's (NIOSH's) Fire Fighter Fatality Investigation and Prevention Program (http://www.cdc.gov/niosh/fire/) is the leading resource in the world for investigations and focused reports on firefighter fatality incidents in the United States of America.

Figure 9. Fatalities by Nature of Fatal Injury (2010).

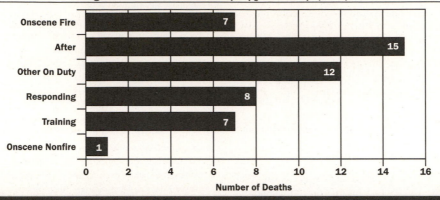

Figure 10. Heart Attacks by Type of Duty (2010).

NIOSH: Preventing Fire Fighter Fatalities Due to Heart Attacks and Other Sudden Cardiovascular Events

[PDF - 1,134 KB]

FIREFIGHTER AGES

Figure 11 shows the percentage distribution of firefighter deaths by age and nature of the fatal injury. Table 11 provides a count of firefighter fatalities by age and the nature of the fatal injury.

Younger firefighters were more likely to have died as a result of traumatic injuries, such as injuries from an apparatus accident or becoming caught or trapped during firefighting operations. Stress-related deaths are rare below the 31 to 35 years of age category and, when they occur, often include underlying medical conditions.

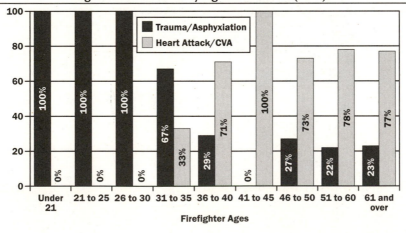

Figure 11. Fatalities by Age and Nature (2010).

Table 11. Firefighter Ages and Nature of Fatal Injury (2010)

Age Range	Heart Attack/CVA/Other	Trauma/Asphyxiation Total
under 21	0	2
21 to 25	0	5
26 to 30	0	4
31 to 35	2	4
36 to 40	5	2
41 to 45	4	0
46 to 50	8	3
51 to 60	18	4
61 and over	20	6

In 2010, there were no teenage firefighters killed while on duty in the United States for the first time since 1993. The two youngest firefighters to die in 2010 were both age 20. Both firefighters were killed in separate incidents involving POV crashes. One firefighter was responding to a fatal motor vehicle accident on a lo cal highway. As he approached the scene, the firefighter, who was not wearing seat restraints, lost control of his pickup truck, struck a guardrail, rolled, and was ejected. The second 20 year old firefighter died while returning from training and riding as a passenger in a car that crashed traveling at high speed while involved in street racing against other firefighters in another vehicle. The accident also claimed the life of another firefighter (driver) and a private citizen riding in another vehicle in the opposite lane of travel who died from injuries nearly 2 weeks later.

The oldest firefighter killed on duty in 2010 was 86. Operating as fire police, he suffered a heart attack after establishing a medivac helispot to evacuate a patient from an EMS incident.

DEATHS BY TIME OF INJURY

The distribution of 2010 firefighter deaths according to the time of day when the fatal injury occurred is illustrated in Figure 12. The time of fatal injury for four firefighters was either unknown or not reported.

Figure 12. Fatalities by Time of Fatal Injury (2010).

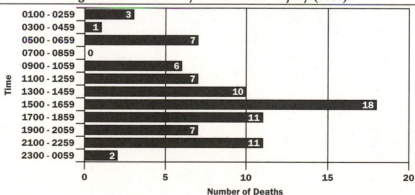

FIREFIGHTER FATALITY INCIDENTS BY MONTH OF YEAR

Figure 13 illustrates the 2010 firefighter fatalities by month of the year.

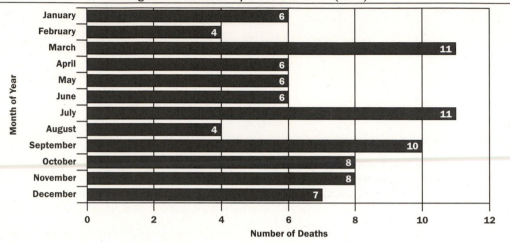

Figure 13. Deaths by Month of Year (2010).

STATE AND REGION

The distribution of firefighter deaths in 2010 by State is shown in Table 12. Firefighters based in 31 States died in 2010.

The highest number of firefighter deaths, based on the location of the fire service organization in 2010, occurred in Illinois with nine deaths. New York and Ohio had the next highest totals of firefighter fatalities in 2010, with eight firefighter deaths each, respectively.

Table 12. Firefighter Fatalities by State by Location of Fire Service* (2010)

State	Fatalities	Percentage
OH	8	9
AR	3	3
NY	8	9
MS	1	1
KY	3	3
ID	1	1
KS	5	5
PA	7	8
DE	1	1
ME	1	1
CT	4	4
WV	1	1

State	Fatalities	Percentage
NC	3	3
AZ	2	2
IL	9	10
MO	2	2
IA	1	1
OK	1	1
WI	1	1
NJ	2	2
GA	2	2
WA	1	1
SC	3	3
TX	2	2
VT	1	1
MA	2	2
VA	3	3
CA	3	3
IN	4	4
TN	1	1
MI	1	1
OH	8	9

*This list attributes the deaths according to the State in which the fire department or unit is based, as opposed to the State in which the death occurred. They are listed by those States for statistical purposes and for the National Fallen Firefighters Memorial at the National Emergency Training Center (NETC). Due to rounding, percentage totals may not add up to 100.

Figure 14. Firefighter Fatalities by Region (2010).

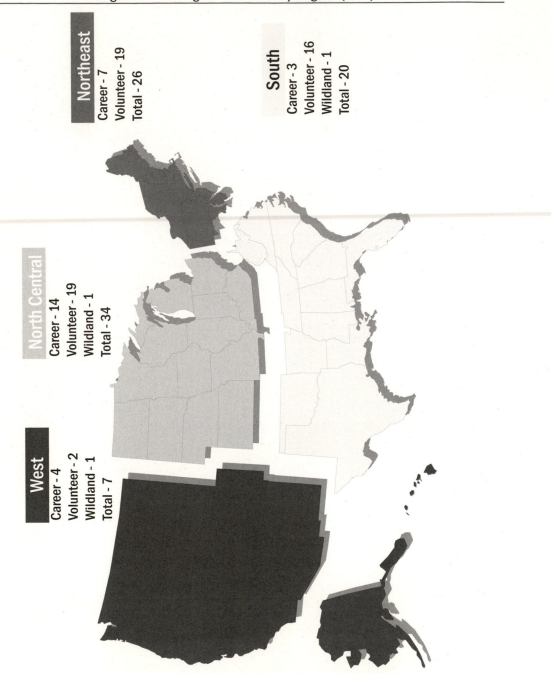

Northeast
Career - 7
Volunteer - 19
Total - 26

South
Career - 3
Volunteer - 16
Wildland - 1
Total - 20

North Central
Career - 14
Volunteer - 19
Wildland - 1
Total - 34

West
Career - 4
Volunteer - 2
Wildland - 1
Total - 7

Figure 15. Onduty Firefighter Fatalities 2010 by Fire Department Location.

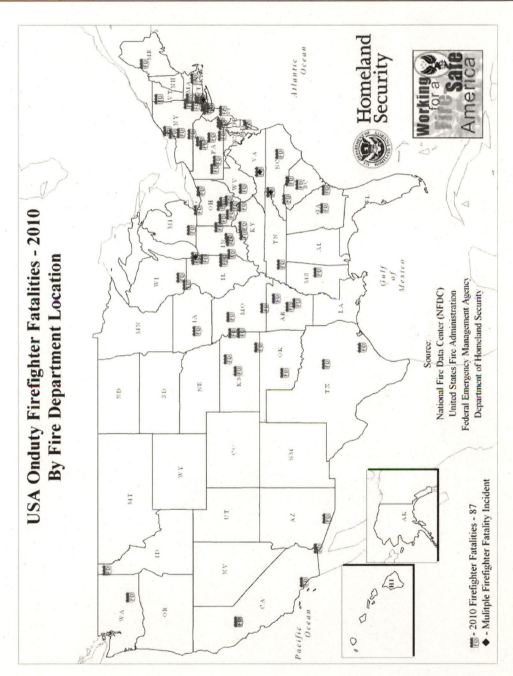

Figure 16. Onduty Firefighter Fatalities 2010 by Incident Location.

ANALYSIS OF URBAN/RURAL/SUBURBAN PATTERNS IN FIREFIGHTER FATALITIES

The U.S. Census Bureau defines "urban" as a place having a population of at least 2,500 or lying within a designated urban area. "Rural" is defined as any community that is not urban. "Suburban" is not a census term but may be taken to refer to any place, urban or rural, that lies within a metropolitan area defined by the Census Bureau, but not within one of the central cities of that metropolitan area.

Fire department areas of responsibility do not always conform to the boundaries used by the Census Bureau. For example, fire departments organized by counties or special fire protection districts may have both urban and rural coverage areas. In such cases, where it may not be possible to characterize the entire coverage area of the fire department as rural or urban, firefighter deaths were listed as urban or rural based on the particular community or location in which the fatality occurred.

The following patterns were found for 2010 firefighter fatalities. These statistics are based on answers from the fire departments and, when no data from the departments were available, the data were based upon population and area served as reported by the fire departments.

Table 13. Firefighter Deaths by Coverage Area Type (2010)

	Urban/Suburban	Rural	Total
Firefighter Deaths	40	47	87

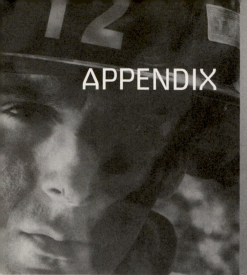

APPENDIX B

Reprint of the Document Issued by the International Association of Fire Chiefs and the International Association of Fire Fighters in 1998 Regarding the Implementation of the Two-in/Two-Out Rule

---

UNITED STATES DEPARTMENT OF LABOR OCCUPATIONAL SAFETY AND HEALTH ADMINISTRATION FIREFIGHTERS' TWO-IN/TWO-OUT REGULATION

The federal Occupational Safety and Health Administration (OSHA) recently issued a revised standard regarding respiratory protection. Among other changes, the regulation now requires that interior structural firefighting procedures provide for at least two firefighters inside the structure. Two firefighters inside the structure must have direct visual or voice contact between each other and direct voice or radio contact with firefighters outside the structure. This section has been dubbed the firefighters' two-in/two-out regulation. The International Association of Fire Fighters and the International Association of Fire Chiefs are providing the following questions and answers to assist you in understanding the section of the regulation related to interior structural firefighting.

1. What is the federal OSHA Respiratory Protection Standard?

In 1971, federal OSHA adopted a respiratory protection standard requiring employers to establish and maintain a respiratory protection program for their respiratory-wearing employees. The revised standard strengthens some requirements and eliminates duplicative requirements in other OSHA health standards.

The standard specifically addresses the use of respirators in immediately dangerous to life or health (IDLH) atmospheres, including interior structural firefighting.

OSHA defines structures that are involved in fire beyond the incipient stage as IDLH atmospheres. In these atmospheres, OSHA requires that personnel use self-contained breathing apparatus (SCBA), that a minimum of two firefighters work as a team inside the structure, and that a minimum of two firefighters be on standby outside the structure to provide assistance or perform rescue.

2. Why is this standard important to firefighters?

This standard, with its two-in/two-out provision, may be one of the most important safety advances for firefighters in this decade. Too many firefighters have died because of insufficient accountability and poor communications. The standard addresses both and leaves no doubt that two-in/two-out requirements must be followed for firefighter safety and compliance with the law.

3. Which firefighters are covered by the regulations?

The federal OSHA standard applies to all private sector workers engaged in firefighting activities through industrial fire brigades, private incorporated fire companies (including the "employees" of incorporated volunteer companies and private fire departments contracting to public jurisdictions) and federal firefighters. In 23 states and 2 territories, the state, not the federal government, has responsibility for enforcing worker health and safety regulations. These "state plan" states have earned the approval of federal OSHA to implement their own enforcement programs. These states must establish and maintain occupational safety and health programs for all public employees that are as effective as the programs for private sector employees. In addition, state safety and health regulations must be at least as stringent as federal OSHA regulations. Federal OSHA has no direct enforcement authority over state and local governments in states that do not have state OSHA plans.

All professional career firefighters, whether state, county, or municipal, in any of the states or territories where an OSHA plan agreement is in effect, have the protection of all federal OSHA health and safety standards, including the new respirator standard and its requirements for firefighting operations. The following states have OSHA-approved plans and must enforce the two-in/two-out provision for all fire departments.

Alaska	Kentucky	North Carolina	Virginia
Arizona	Maryland	Oregon	Virgin Islands
California	Michigan	Puerto Rico	Washington
Connecticut	Minnesota	South Carolina	Wyoming
Hawaii	Nevada	Tennessee	
Indiana	New Mexico	Utah	
Iowa	New York	Vermont	

A number of other states have adopted, by reference, federal OSHA regulations for public employee fire fighters. These states include Florida, Illinois, and Oklahoma. In these states, the regulations carry the force of state law.

Additionally, a number of states have adopted NFPA standards, including *NFPA 1500, Standard for Fire Department Occupational Safety and Health Program*. The 1997 edition of *NFPA 1500* now includes requirements corresponding to OSHA's respiratory protection regulation. Since the NFPA is a private consensus standards organization, its recommendations are preempted by OSHA regulations that are more stringent. In other words, the OSHA regulations are the minimum requirement where they are legally applicable. There is nothing in federal regulations that "deem compliance" with any consensus standards, including NFPA standards, if the consensus standards are less stringent.

It is unfortunate that all U.S. and Canadian firefighters are not covered by the OSHA respiratory protection standard. However, we must consider the two-in/two-out requirements to be the minimum acceptable standard for safe fireground operations for all firefighters when self-contained breathing apparatus is used.

4. When are two-in/two-out procedures required for firefighters?

OSHA states that "once fire fighters begin the interior attack on an interior structural fire, the atmosphere is assumed to be IDLH and paragraph 29 CFR 1910.134(g)(4) [two-in/two-out] applies." OSHA defines interior structural fire fighting "as the physical activity of fire suppression, rescue or both inside of buildings or enclosed structures which are involved in a fire situation beyond the incipient stage." OSHA further defines an incipient stage fire in 29 CFR 1910.155(c)(26) as a "fire which is in the initial or beginning stage and which can be controlled or extinguished by portable fire extinguishers, Class II standpipe or small hose systems without the need for protective clothing or breathing apparatus." Any structural fire beyond incipient stage is considered to be an IDLH atmosphere by OSHA.

5. What respiratory protection is required for interior structural firefighting?

OSHA requires that all firefighters engaged in interior structural firefighting must wear SCBAs. SCBAs must be NIOSH-certified, positive pressure, with a minimum duration of 30 minutes. [29 CFR 1910.156(f)(1)(ii)] and [29 CFR 1910.134(g)(4)(iii)]

6. Are all firefighters performing interior structural firefighting operations required to operate in a buddy system with two or more personnel?

Yes. OSHA clearly requires that all workers engaged in interior structural firefighting operations beyond the incipient stage use SCBA and work in teams of two or more. [29 CFR 1910.134(g)(4)(i)]

7. Are firefighters in the interior of the structure required to be in direct contact with one another?

Yes. Firefighters operating in the interior of the structure must operate in a buddy system and maintain voice or visual contact with one another at all times. This assists in assuring accountability within the team. [29 CFR 1910.134(g)(4)(i)]

8. Can radios or other means of electronic contact be substituted for visual or voice contact, allowing firefighters in an interior structural fire to separate from their "buddy" or "buddies"?

No. Due to the potential of mechanical failure or reception failure of electronic communication devices, radio contact is not acceptable to replace visual or voice contact between the members of the buddy system team. Also, the individual needing rescue may not be physically able to operate an electronic device to alert other members of the interior team that assistance is needed.

Radios can and should be used for communications on the fireground, including communications between the interior firefighter team(s) and exterior firefighters. They cannot, however, be the sole tool for accounting for one's partner in the interior of a structural fire. [29 CFR 1910.134(g)(4)(i)] [29 CFR 1910.134(g)(3)(ii)]

9. Are firefighters required to be present outside the structural fire prior to a team entering and during the team's work in the hazard area?

Yes. OSHA requires at least one team of two or more properly equipped and trained firefighters be present outside the structure before any team(s) of firefighters enter the structural fire. This requirement is intended to ensure that the team outside the structure has the training, clothing, and equipment to protect themselves and, if necessary, safely and effectively rescue firefighters inside the structure. For high-rise operations, the team(s) would be staged below the IDLH atmosphere. [29 CFR 1910.134(g)(3)(iii)]

10. Do these regulations mean that, at a minimum, four individuals are required, that is, two individuals working as a team in the interior of the structural fire and two individuals outside the structure for assistance or rescue?

Yes. OSHA requires that a minimum of two individuals, operating as a team in direct voice or visual contact, conduct interior firefighting operations utilizing SCBA. In addition, a minimum of two individuals who are properly equipped and trained must be positioned outside the IDLH atmosphere, account for the interior team(s), and remain capable of rapid rescue of the interior team. The outside personnel must at all times account for and be available to assist or rescue members of the interior team. [29 CFR 1910.134(g)(4)]

11. Does OSHA permit the two individuals outside the hazard area to be engaged in other activities, such as incident command or fire apparatus operation (for example, pump or aerial operators)?

OSHA requires that one of the two outside person's function is to account for and, if necessary, initiate a firefighter rescue. Aside from this individual dedicated to tracking interior personnel, the other designated person(s) is permitted to take on other roles, such as incident commander in charge of the emergency incident, safety officer, or equipment operator. However, the other designated outside person(s) cannot be assigned tasks that are critical to the safety and health of any other employee working at the incident.

Any task that the outside firefighter(s) performs while in standby rescue status must not interfere with the responsibility to account for those individuals in the hazard area. Any task, evolution, duty, or function being performed by the standby individual(s) must be such that the work can be abandoned, without placing any employee at additional risk, if rescue or other assistance is needed. [29 CFR 1910.134(g)(4) (Note 1)]

12. If a rescue operation is necessary, must the buddy system be maintained while entering the interior structural fire?

Yes. Any entry into an interior structural fire beyond the incipient stage, regardless of the reason, must be made in teams of two or more individuals. [29 CFR 1910.134(g)(4)(i)]

13. Do the regulations require two individuals outside for each team of individuals operating in the interior of a structural fire?

The regulations do not require a separate two-out team for each team operating in the structure. However, if the incident escalates, if accountability cannot be properly maintained from a single exposure, or if rapid rescue becomes infeasible, additional outside crews must be added. For example, if the involved structure is large enough to require entry at different locations or levels, additional two-out teams would be required. [29 CFR 1910.134(g)(4)]

14. If four firefighters are on the scene of an interior structural fire, is it permissible to enter the structure with a team of two?

OSHA's respiratory protection standard is not about counting heads. Rather, it dictates functions of firefighters prior to an interior attack. The entry team must consist of at least two individuals. Of the two firefighters outside, one must perform accountability functions and be immediately available for firefighter rescue. As explained previously, the other may perform other tasks, as long as those tasks do not interfere with the accountability functions and can be abandoned to perform firefighter rescue. Depending on the operating procedures of the fire department, more than four individuals may be required. [29 CFR 1910.134(g)(4)(i)]

15. Does OSHA recognize any exceptions to this regulation?

OSHA regulations recognize deviations to regulations in an emergency operation in which immediate action is necessary to save a life. For fire department employers, initial attack operations must be organized to ensure that adequate personnel are at the emergency scene prior to any interior firefighting at a structural fire. If initial attack personnel find a known life-hazard situation in which immediate action could prevent the loss of life, deviation from the two-in/two-out standard may be permitted, as an exception to the fire department's organizational plan.

However, such deviations from the regulations must be exceptions and not de facto standard practices. In fact, OSHA may still issue de minimis citations for such deviations from the standard, meaning that the citation will not require monetary penalties or corrective action. The exception is for a known life rescue only, not for standard search and rescue activities. When the exception becomes the practice, OSHA citations are authorized. [29 CFR 1910.134(g)(4)(Note 2)]

16. Does OSHA require employer notification prior to any rescue by the outside personnel?

Yes. OSHA requires the fire department or fire department designee (i.e., incident commander) be notified prior to any rescue of firefighters operating in an IDLH atmosphere. The fire department would have to provide any additional assistance appropriate to the emergency, including the notification of on-scene personnel and incoming units. Additionally, any such actions taken in accordance with the "exception" provision should be thoroughly investigated by the fire department with a written report submitted to the fire chief. [29 CFR 1910.134(g)(3)(iv)]

17. How do the regulations affect firefighters entering a hazardous environment that is not an interior structural fire?

Firefighters must adhere to the two-in/two-out regulations for other emergency response operations in any IDLH, potential IDLH, or unknown atmosphere. OSHA permits one standby person only in those IDLH environments in fixed workplaces, not fire emergency situations. Such sites, in normal operating conditions, contain only hazards that are known, well characterized, and well controlled. [29 CFR 1910.120(q)(3)(vi)]

18. When is the new regulation effective?

The revised OSHA respiratory protection standard was released by the Department of Labor and published in the *Federal Register* on January 8, 1998. It is effective on April 8, 1998.

"State Plan" states have six months from the release date to implement and enforce the new regulations.

Until the April 8 effective date, earlier requirements for two in/two out are in effect. The formal interpretation and compliance memo issued by James W. Stanley, Deputy Assistant Secretary of Labor, on May 1, 1995 and the compliance memo issued by Assistant Secretary of Labor Joe Dear on July 30, 1996 establish that OSHA interprets the earlier 1971 regulation as requiring two in/two out. [29 CFR 1910.134(n)(1)]

19. How does a fire department demonstrate compliance with the regulations?

Fire departments must develop and implement standard operating procedures addressing fireground operations and the two-in/two-out procedures to demonstrate compliance. Fire department training programs must ensure that firefighters understand and implement appropriate two-in/two-out procedures. [29 CFR 1910.134(c)]

20. What can be done if the fire department does not comply?

Federal OSHA and approved state plan states must "... assure so far as possible every working man and woman in the Nation safe and healthful working conditions." To ensure such protection, federal OSHA and states with approved state plans are authorized to enforce safety and health standards. These agencies must investigate complaints and conduct inspections to make sure that specific standards are met and that the workplace is generally free from recognized hazards likely to cause death or serious physical harm.

Federal OSHA and state occupational safety and health agencies must investigate written complaints signed by current employees or their representatives regarding hazards that threaten serious physical harm to workers. By law, federal and state OSHA agencies do not reveal the name of the person filing the complaint, if he or she so requests. Complaints regarding imminent danger are investigated even if they are unsigned or anonymous. For all other complaints (from other than a current employee, or unsigned, or anonymous), the agency may send a letter to the employer describing the complaint and requesting a response. It is important that an OSHA (either federal or state) complaint be in writing.

When an OSHA inspector arrives, he or she displays official credentials and asks to see the employer. The inspector explains the nature of the visit, the scope of the inspection, and applicable standards. A copy of any employee complaint (edited, if requested, to conceal the employee's identity) is available to the employer. An employer representative may accompany the inspector during the inspection. An authorized representative of the employees, if any, also has the right to participate in the inspection. The inspector may review records, collect information and view work sites. The inspector may also interview employees in private for additional information. Federal law prohibits discrimination in any form by employers against workers because of anything that workers say or show the inspector during the inspection or for any other OSHA-protected safety-related activity.

Investigations of imminent danger situations have top priority. An imminent danger is a hazard that could cause death or serious physical harm immediately, or before the danger can be eliminated through normal enforcement procedures. Because of the hazardous and unpredictable nature of the fireground, a fire department's failure to comply with the two-in/two-out requirements creates an imminent danger and the agency receiving a related complaint must provide an immediate response. If inspectors find imminent danger conditions, they will ask for immediate voluntary correction of the hazard by the employer or removal of endangered employees from the area. If an employer fails to do so, federal OSHA can go to federal district court to force the employer to comply. State occupational safety and health agencies rely on state courts for similar authority.

Federal and state OSHA agencies are required by law to issue citations for violations for safety and health standards. The agencies are not permitted to issue warnings. Citations include a description of the violation, the proposed penalty (if any), and the date by which the hazard must be corrected. Citations must be posted in the workplace to inform employees about the violation and the corrective action. [29 CFR 1903.3(a)]

It is important for labor and management to know that this regulation can also be used as evidence of industry standards and feasibility in arbitration and grievance hearings on firefighter safety, as well as in other civil or criminal legal proceedings involving injury or death in which the cause can be attributed to employer failure to implement two-in/two-out procedures. Regardless of OSHA's enforcement authority, this federal regulation links fireground operations with firefighter safety.

21. What can be done if a firefighter does not comply with fire department operating procedures for two in/two out?

Fire departments must amend any existing policies and operational procedures to address the two-in/two-out regulations and develop clear protocols and reporting procedures for deviations from these fire department policies and procedures. Any individual violating this safety regulation should face appropriate departmental action.

22. How can I obtain additional information regarding the OSHA respirator standard and the two-in/two-out provision?

Affiliates of the International Association of Fire Fighters may contact:

International Association of Fire Fighters
Department of Occupational Health and Safety
1750 New York Avenue, NW
Washington, DC 20006
202-737-8484
202-737-8418 (FAX)

Members of the International Association of Fire Chiefs may contact:

International Association of Fire Chiefs
4025 Fair Ridge Drive
Fairfax, VA 22033-2868
703-273-0911
703-273-9363 (FAX)

DOES THE TWO-IN/TWO-OUT REGULATION APPLY TO VOLUNTEER FIRE DEPARTMENTS AND VOLUNTEER FIREFIGHTERS?

Yes, it is the opinion of the IAFC Volunteer Chief Officers Section (VCOS) that this regulation is applicable in all situations in which interior firefighting operations are being conducted. VCOS believes it is applicable to the vast majority of volunteer fire departments in the United States.

The VCOS supported the concept of two in/two out when it was originally proposed and VCOS still supports these new regulations since they deal directly with the safety and well-being of our firefighters.

While OSHA has stated that volunteer fire departments will not be affected by the regulations, VCOS believes that this will not be the case since 25 states align with federal OSHA. The standard also applies to private incorporated fire companies including the "employees" of incorporated volunteer companies. In addition, other non-OSHA states are giving the regulations consideration under the Environmental Protection Agency, the federal entity that typically takes the lead in non-OSHA states.

The application of the two-in/two-out rule can be argued by volunteer fire departments in many states. The reality is that two in/two out is the right thing to do for the safety of our firefighters and each volunteer fire department should seek to implement new standard operating guidelines for two in/two out.

For further information on the application of two in/two out to volunteers, contact:

Volunteer Chief Officers Section
International Association of Fire Chiefs
4025 Fair Ridge Drive
Fairfax, VA 22033-2868
703-273-0911 703-273-9363 (FAX)

APPENDIX C

Firefighter Life Safety Summit Initial Report

FIREFIGHTER LIFE SAFETY SUMMIT INITIAL REPORT

Courtesy of Federal Emergency Management Agency (FEMA)

April 14, 2004

PART I

An unprecedented gathering of the leadership of the American fire service occurred on March 10th and 11th, 2004, when more than 200 individuals assembled in Tampa to focus on the troubling question of how to prevent line-of-duty deaths. Every year approximately 100 firefighters lose their lives in the line of duty in the United States; about one every 80 hours. The first ever National Fire Fighter Life Safety Summit was convened to bring the leadership of the fire service together for two days to focus all of their attention on this one critical concern. Every identifiable segment of the fire service was represented and participated in the process.

The National Fallen Firefighters Foundation hosted the Summit as the first step in a major campaign. In cooperation with the United States Fire Administration, the Foundation has established the objectives of reducing the fatality rate by 25% within 5 years and by 50% within 10 years. The purpose of the Summit was to produce an agenda of initiatives that must be addressed to reach those milestones and to gain the commitment of the fire service leadership to support and work toward their accomplishment.

The summit marks a significant milestone, because it is the first time that a major gathering has been organized to unite all segments of the fire service behind the common goal of reducing firefighter deaths. It provided an opportunity for all of the participants to focus on the problems, jointly identify the most important issues, agree upon a set of key initiatives, and develop the commitments and coalitions that are essential to move forward with their implementation.

Every individual who came to Tampa was already personally committed to the mission of keeping firefighters alive and all of the organizations that were represented were already on record as supporting the goal of reducing line-of-duty deaths. The Summit was designed to produce a single combined agenda for change that all of the participants, individuals and organizations could agree to support and promote. The product of their concentrated effort in Tampa will provide the foundation for a joint strategy and combined effort that will be essential to produce the desired results over the next ten years.

The Summit produced a set of initiatives that may well be regarded as radical today, however it is significant to recognize that nothing new was invented or discovered in Tampa. All of the initiatives that emerged were based on information and fundamental truths that were known long before the invitations to the Summit were issued. The gathering simply provided a forum at which those issues could be discussed openly and freely on their own merits. Some of the policies that were identified are likely to cause discomfort and controversy, however there is no arguing with the fact that the assembled leadership, who came from all segments of the fire service, concluded that these initiatives are essential to keep firefighters from dying unnecessarily.

This is the first step along a path that will require a huge commitment of energy and resources over several years. Some of the initiatives that were agreed upon will involve radical changes for the fire service. Any revolutionary movement requires committed and unwavering leadership to bring about this type of major change. The core of that leadership will come from the Summit participants who helped to shape the agenda and identify the strategies that will have to be implemented. The invited participants included key individuals who are widely recognized for their influence and leadership, some attending on their own and some as representatives of organizations that represent different sectors of the fire service. In the normal course of events these organizations often disagree on particular issues and priorities, however in Tampa the only issue on the agenda was how to keep firefighters alive and there was a very broad consensus on the efforts that are needed to accomplish that goal. As the initiatives are advanced over the next several years, the fire service will see an example of what committed leaders can accomplish when they agree to work together for an important cause.

Process

The Summit opened with an immediate emphasis on the need to take bold action to change perceptions and expectations in the fire service. The strongest words of inspiration came from NFFF Board Member Vina Drennan, who clearly reminded everyone of the pain that is felt by surviving family members and shared throughout the fire service whenever a member is killed, particularly when a life is lost in circumstances that could have been prevented. A review of historical data and statistical trends in line-of-duty deaths was presented to help the participants appreciate the range of problems and issues that must be addressed to achieve a significant reduction in fatalities.

The majority of the effort during the Summit took place in six discussion groups that focused their attention on specific domains. Groups were assembled to address:

- Structural firefighting
- Wildland firefighting
- Training and research
- Vehicle operations
- Health-wellness-fitness
- Reduction of emergency incidents and risks.

The groups were asked to produce a set of initiatives that should be undertaken to reduce line-of-duty deaths within their assigned domains. Each group was assigned two co-facilitators to lead the discussions, as well as a staff assistant to fully document the discussions. Within each of the domains the participants were asked to consider:

- Education and awareness issues
- Standards and regulations
- Specific research and technology issues
- Psychological barriers
- Leadership and personal/professional responsibility issues

The groups reported their recommendations back to the full assembly, which then produced a single consolidated set of key initiatives and implementation strategies as the product of the Summit.

The consolidated list included 16 individual indicatives:

1. Define and advocate the need for a cultural change within the fire service relating to safety incorporating leadership, management, supervision, accountability, and personal responsibility.
2. Enhance the personal and organizational accountability for health and safety throughout the fire service.
3. Focus greater attention on the integration of risk management with incident management at all levels, including strategic, tactical, and planning responsibilities.
4. All firefighters must be empowered to stop unsafe practices.
5. Develop and implement national standards for training, qualifications, and certification (including regular recertification) that are equally applicable to all firefighters based on the duties they are expected to perform.
6. Develop and implement national medical and physical fitness standards that are equally applicable to all firefighters, based on the duties they are expected to perform.
7. Create a national research agenda and data collection system that relates to the initiatives.
8. Utilize available technology wherever it can produce higher levels of health and safety.
9. Thoroughly investigate all firefighter fatalities, injuries, and near misses.
10. Grant programs should support the implementation of safe practices and/or mandate safe practices as an eligibility requirement.
11. National standards for emergency response policies and procedures should be developed and championed.
12. National protocols for response to violent incidents should be developed and championed.
13. Firefighters and their families must have access to counseling and psychological support.
14. Public education must receive more resources and be championed as a critical fire and life safety program.
15. Advocacy must be strengthened for the enforcement of codes and the installation of home fire sprinklers.
16. Safety must be a primary consideration in the design of apparatus and equipment.

Suggested Role for the National Fallen Firefighters Foundation

While developing the agenda of initiatives to reduce firefighter fatalities, the delegates were also asked to define an appropriate role for The National Fallen Firefighters Foundation in working toward the implementation of the Summit recommendations. The Foundation is committed to this mission and to taking on roles and responsibilities that it can perform more effectively than any other group. The suggestions included a list of potential approaches that could be adopted by the Foundation:

- Coalition building
- Provide advocacy for the issues
- Activism in the standards-making process
- Political and apolitical activism
- Partnerships and fundraising
- Catalyst for change
- Serve as a clearinghouse for information and data
- Develop model programs and demonstration projects
- Provide technical assistance
- Work with NIOSH on investigations
- Timely assistance to local jurisdictions
- Provide recognition for achievements and contributions
- Review performance and provide progress reports
- Public awareness and communications

The Foundation's Board of Directors will consider all of the suggested roles in the coming months and determine the ability of the Foundation to undertake each activity or project. The Board has already committed the Foundation to assume a set of responsibilities, beginning with organizing the Summit and serving as the key communicator and advocacy group for the initiatives that were developed.

In the immediate future the Foundation will join forces with the United States Fire Administration to document, publish and distribute the initiatives. The Foundation will also work with USFA to report on the progress that is achieved in implementing each of the initiatives. The Foundation has already begun to seek partners and ambassadors to move the initiatives forward and scheduled a strategy meeting to establish the immediate, mid-range and long-range priorities for a 10-year campaign. That effort should be extremely visible in the coming months.

PART II

Background discussion related to the initiatives, summarized from the Summit

Cultural Change

The most fundamental issue that was agreed upon by the Summit participants is the need for the fire service in the United States to change the culture of accepting the loss of firefighters as a normal way of doing business. This concept was reflected in several different statements that were produced by the individual discussion groups. The Summit participants unanimously declared that the time has come to change our culture and our expectations.

Within the fire service we all feel the pain with the loss of each individual firefighter, but we have come to accept the loss of more than 100 firefighters each year as a standard expectation. As long as we continue to accept this loss, we can avoid or delay making the radical and uncomfortable adjustments that will be necessary to change the outcomes. We have to convince everyone in the fire service that a line of duty death is not a standard expectation or an acceptable outcome.

Personal and Organizational Accountability

The essential cultural change has to begin with accepting personal and organizational accountability for health and safety. Every individual within the fire service has to accept a personal responsibility for health, wellness, fitness for duty, skills development, basic competencies and adherence to safe practices. The leaders and members of every fire department and every fire service organization must be accountable for the safety of their members, collectively and individually. In addition the members must be accountable to each other.

The most important and fundamental decisions relating to firefighter health and safety are made by individuals, from the top of the organizational chart to the bottom. Irresponsible behavior cannot be tolerated at any level and no external influence can overpower a failure to accept personal responsibility. The managers, supervisors and leaders within the fire service must instill and reinforce these values until they become an integral component of the culture.

Incident Management and Risk Management

This initiative incorporates a range of components that relate to our ability to safely conduct emergency operations in a high-risk environment. There is no question that fire fighters are expected to work in an environment that is inherently dangerous, however the risks and most of the specific dangers are well known. The most common causes of firefighter deaths are widely recognized, along with the situations where they are most likely to occur. We have to recognize and manage the risks that apply to each situation. The essence of professionalism in the fire service is the ability to function safely and effectively within that dangerous environment. We will never be able to eliminate all of the risks, but we can be very well prepared to face most of them.

Firefighters at every level must be properly trained, equipped, organized and directed to perform their duties safely and skillfully. There must be a comprehensive structured system in place to manage incidents and risks. Company officers must be trained to supervise operations and incident commanders must be trained to manage incidents according to standard principles and practices.

Firefighters must be prepared to function competently in a wide range of situations, including critical events that can involve unanticipated dangers and immediate risks to their own survival. Several areas were identified for special emphasis, including mayday and rapid intervention procedures, air management and preventing disorientation in zero-visibility conditions.

Risk management involves identifying the situations where predictable risks are likely to be encountered and making decisions that will reduce, eliminate or avoid them.

Realistic risk management applies at every level within the fire service, from the decisions made by individual firefighters and company officers to the actions of incident commanders and senior officers who have specific responsibilities for evaluating and managing risks. We fail to act professionally when we recognize a risk and choose to do nothing about it.

The willingness of firefighters to risk their own lives to save others must never be used as an excuse to take unnecessary risks. Firefighters are highly respected for being willing to risk their own lives to save others, but that cannot justify taking unnecessary risks in situations where there is no one to save and nothing to be gained. In too many cases firefighters lose their lives while trying to save property that is already lost or to rescue victims who are already dead. While these efforts are valiant, they are also futile. Individual firefighters who take unnecessary risks, or fail to follow standard safety practices, endanger their own lives as well as the lives of other fire fighters who are depending on them or who might have to try to rescue them.

Right to Stop Unsafe Procedures

The Summit participants identified the fundamental right and responsibility of firefighters to stop unsafe procedures as a key issue. To many members of the fire service, who have been indoctrinated with a traditional sense of unquestioning discipline, this could be an uncomfortable concept. The underlying principle is that an individual who recognizes an unsafe situation must take action to prevent an accident from occurring. Under this operational concept, any firefighter who believes that a situation is unsafe, or could be unsafe, has both the right and the responsibility to stop the action while an evaluation is made.

The justification for this policy in non-emergency activities, particularly training situations, is easily understood. The application of the same concept to emergency operations could be more difficult for some individuals to accept. In too many cases the investigation of a fatal accident determines that an unsafe situation was recognized, but no action was taken to change or reconsider the plan. There are very few situations, even during emergency operations, where a brief hesitation to re-evaluate a potentially dangerous plan of action would have serious negative consequences. On the contrary, experience has shown that many lives could have been saved by taking a few extra seconds to stop and think.

This policy does not mean that no action can be taken at an emergency scene that exceeds the comfort level of any individual. The obligation attached to the policy is to pause long enough to determine if it is reasonably safe to continue. In many cases an officer will have to make a very rapid assessment of the situation and decide whether to continue or change the plan.

During emergency incidents there is one key question that we need to ask ourselves at regular intervals: "Are the results we are trying to accomplish worth the risks we are taking with our people?" The answer to this question should dictate actions by commanders and firefighters.

Mandatory National Standards for Training and Qualifications

The Summit participants reached the conclusion that the time has come to apply a mandatory national uniform system of training and qualification standards for all firefighters. This system would establish mandatory training, education and performance requirements, based on the duties an individual is expected to perform, regardless of their status within the fire service or the type of organization. The roles and responsibilities of firefighters at different levels and in different operating environments would be clearly defined.

The basic system of professional qualifications standards already exists, however their current applicability depends on state and local jurisdictions, individual organizations, seniority, whether a firefighter is a career, part-time or volunteer member of the fire service, and many other factors. The summit delegates agreed that we must move toward a system where the same standards would be applied to anyone who performs a given role within the fire service. The applicable standards must be appropriate and realistic for the functions the individual is expected to perform, however there should not be standards that apply to some individuals and not to others.

In reference to the existing professional qualification and certification systems, the need for periodic recertification was identified as a priority. Today, in many cases, an individual can be certified at a particular level and retain that certification for life, with no requirements for continuing education, refresher classes, performance testing or skills evaluation. The established systems for emergency medical practitioner certification were suggested as a model.

Mandatory Medical and Physical Fitness Standards

Mandatory requirements for medical examinations and physical fitness standards should also be implemented for all firefighters, based on the duties and functions they are expected to perform. Medical surveillance should be increased before strenuous physical activity and during activities where firefighters are expected to operate at extreme levels of exertion and endurance.

An increased emphasis on health, wellness and fitness is essential to reduce the number of deaths resulting from heart attacks and other cardiovascular causes. Statistics suggest that the most significant reductions in line of duty deaths are likely to be achieved through increased medical surveillance and physical fitness programs. The need for improvements in this area is most pronounced in the volunteer fire service, where the rate of fatalities due to heart attacks and other cardiovascular causes is now much higher than within the career service. This is a reversal of the situation that existed twenty years ago, when there were more cardiovascular deaths among career firefighters than volunteers.

National Research Agenda and Data Collection

The discussion of on-going research efforts as well as priorities for additional research projects pointed to the need for a national research agenda and budget for the fire service. Several different areas were listed as operational research priorities, from building construction to communications systems, leadership, management practices, and decision-making. A separate research agenda relating to the full spectrum of health, wellness and physical fitness issues was also identified, including psychological and physiological stress, cardiovascular function, oncology and biofeedback. The need for more comprehensive data collection and sharing to identify problem areas and support research was also recognized.

The researchers who were present described their current projects, as well as asking for guidance on the most pressing areas to explore. Several research projects are being conducted by different organizations on a variety of subjects relating to health and safety, however most of their funding and support come from sources outside the fire service and the projects are often influenced by other priorities and agendas. The fire service needs a process to identify its own research priorities and coordinate efforts, as well as a dedicated source of funding that can be allocated to the most important research projects. The delegates also recognized the challenge of applying many of the current research findings to real world situations.

The fire service is benefiting from crossover adaptations of technologies that were developed for other purposes; however only limited funding is available to develop new technology specifically for fire service applications. The delegates emphasized the need to make the most use of technological advancements that are available to reduce the risks of emergency operations and training exercises. Cost is often mentioned as a barrier that keeps fire departments from adopting technological improvements that could improve the safety of operations. If a technological solution is available to eliminate a known risk, the cost must be considered in relation to the consequences of not making the investment to protect the lives of firefighters.

Fatality Investigations

The need for more consistent and comprehensive investigations and data collection to analyze the causes of fire fighter fatalities was also identified as an important priority. The delegates noted that the existing NIOSH fatality investigation program needs to be expanded and that every fatality should be thoroughly investigated and documented by a team of investigators who are qualified to examine all of the pertinent factors. The same type of investigation should be conducted for serious injuries and near-miss incidents to focus on preventing future occurrences. The need for autopsy results based on a standard protocol for every line of duty death was reinforced.

Safety in Apparatus and Equipment Design

Improvements in the design and construction of apparatus and equipment are needed to address a long list of concerns. The areas discussed ranged from breathing apparatus improvements to reducing the risk of rollover accidents involving tanker apparatus. Several on-going research and development projects were described and the manufacturers and suppliers who were present noted numerous suggestions for potential improvements.

Several safety issues relating to the danger of firefighters being struck by vehicles while operating on roadways were discussed. A comprehensive approach involving apparatus positioning, emergency lighting, warning signs and traffic control devices, high visibility protective clothing, coordination with police agencies and public education was identified as a high priority for nationwide training and implementation.

Incentives and Grants Tied to Safe Practices

Several discussion groups identified the need to create a direct link between reinforcing appropriate safety policies and practices and the availability of state and federal grant funds for fire departments. There was unanimous support for the basic concept of using grant funds to leverage improvements in heath and safety programs. The suggestions ranged from making special incentive grants to fire departments in order to implement safe practices to making compliance with health and safety standards a condition for receiving funds under existing grant programs.

The Summit participants also supported the idea of officially recognizing achievements and significant improvements in heath and safety. Several individual projects and organizations were suggested as already deserving recognition.

Response Policies

Another need for cultural change was identified in relation to emergency vehicle operations and response policies. The group that examined this area noted that an average of

10 firefighters are killed each year in vehicle accidents while responding to emergency incidents and an even greater number of civilians die in collisions involving responding emergency vehicles.

Many of the emergency response deaths result from excessive speed and unsafe driving, which can be related to the perception that the urgency of the mission justifies an elevated level of risk to the emergency responders and everyone else on the streets. In too many cases the risks that are created en route are greater than the dangers of the situation itself. The cultural change must be based on recognizing that firefighters cannot save lives or property at the scene of an emergency incident unless they arrive safely and there is no justification for causing more harm en route than they can prevent when they arrive.

The need for standard policies governing emergency response was identified, possibly through the development of a new NFPA standard or by adding to the scope of an existing standard. These policies should determine when emergency response is and is not appropriate and include specific policies relating to responding in privately owned vehicles.

This cultural change has to begin with the enforcement of existing safe driving protocols by leaders and supervisors, as well as the mandatory use of seatbelts by all firefighters. The delegates noted that in many cases firefighters do not use seatbelts that are provided in their vehicles, in spite of NFPA standards, departmental regulations and state laws. The failure to enforce and to follow these existing and basic safety procedures was highlighted as evidence of the urgent need for cultural change.

The delegates recommended the adoption of a special classification of driver's license for emergency vehicle operators, similar to the existing commercial driver's license program. Instead of providing special exemptions for emergency vehicle operators, regulatory authorities should establish strict training and testing requirements, including medical clearances and periodic review of driving records. The highest standards should be applied to emergency vehicle operators.

Protocol for Violent Incidents

The Summit delegates identified the need for special protocols and policies in several areas, including a protocol for responding to violent incidents. This recognized the increasing exposure of firefighters to violent crimes and situations, up to and including terrorist attacks.

Counseling and Psychological Support

Based largely on experiences related to 9-11 and FDNY, counseling and psychological support programs were identified as high priorities for increased attention. Some of the accepted concepts and programs have been found to be inadequate or counterproductive. A combination of research efforts and large-scale practical experience will be required to make the necessary changes and improvements to these programs.

Public Education

The potential impact of public education on firefighter safety was addressed in a variety of contexts. Public fire and life safety education was identified as a strategy to reduce fires and the resulting level of risk to firefighters. In a similar manner, programs that are designed to prevent injuries and teach citizen CPR will reduce the risks encountered when responding to emergency medical and rescue incidents. Teaching drivers how to react appropriately when they encounter an emergency vehicle or when approaching an incident on a roadway will reduce another component of risk to firefighters.

Fire Prevention Codes and Residential Sprinklers

One of the most productive strategies for reducing the risk of fire fighter fatalities is to reduce the frequency of fires and emergency incidents. A comprehensive effort to increase fire service activism in fire prevention, code development and code enforcement should have a direct impact on reducing the exposure of fire fighters to dangerous situations. Efforts to promote the installation of automatic sprinkler systems in all new residential construction will have a profound impact on future fire rates. Addressing the problem of juvenile fire setters will have a positive impact on firefighter safety as well as general public safety. The delegates endorsed placing an increased emphasis on prevention as a long-term strategy to reduce fire fighter fatalities.

FIREFIGHTER LIFE SAFETY INITIATIVES

1. Define and advocate the need for a cultural change within the fire service relating to safety incorporating leadership, management, supervision, accountability, and personal responsibility.

2. Enhance the personal and organizational accountability for health and safety throughout the fire service.

3. Focus greater attention on the integration of risk management with incident management at all levels, including strategic, tactical, and planning responsibilities.

4. All firefighters must be empowered to stop unsafe practices.

5. Develop and implement national standards for training, qualifications, and certification (including regular recertification) that are equally applicable to all firefighters based on the duties they are expected to perform.

6. Develop and implement national medical and physical fitness standards that are equally applicable to all firefighters, based on the duties they are expected to perform.

7. Create a national research agenda and data collection system that relates to the initiatives.

8. Utilize available technology wherever it can produce higher levels of health and safety.

9. Thoroughly investigate all firefighter fatalities, injuries, and near misses.

10. Grant programs should support the implementation of safe practices and/or mandate safe practices as an eligibility requirement.

11. National standards for emergency response policies and procedures should be developed and championed.

12. National protocols for response to violent incidents should be developed and championed.

13. Firefighters and their families must have access to counseling and psychological support.

14. Public education must receive more resources and be championed as a critical fire and life safety program.

15. Advocacy must be strengthened for the enforcement of codes and the installation of home fire sprinklers.

16. Safety must be a primary consideration in the design of apparatus and equipment.

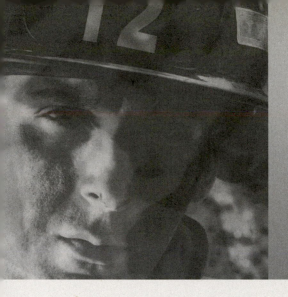

GLOSSARY

A

Aerobic fitness A measurement of the body's ability to perform and utilize oxygen.

B

Back drafts Occur as a result of burning in an oxygen-starved atmosphere. When air is introduced, the superheated gases ignite with enough force to be considered an explosion.

Bloodborne pathogen Disease carried in blood or blood products.

Body composition A measure to show the percentage of fat in the body; there are certain published parameters for what is considered average or normal.

Body substance isolation (BSI) An infection control strategy that considers all body substances potentially infectious and isolates all body substances that might be infectious.

C

Cardiovascular fitness Fitness levels associated with the cardiovascular system, including the heart and circulatory system.

Cause-and-effect diagrams Are causal diagrams that show the causes of a certain event.

Civil disturbances Uprisings of civilians that often lead to hostile acts against law enforcement and emergency responders.

Code of Federal Regulations (CFR) The document that contains all of the federally promulgated regulations for all federal agencies.

Cognitive Skills learned through a mental learning process as opposed to practical learning.

Consensus standards Standards developed by consensus of industry or subject area experts, which are then published and may or may not be adopted locally. Even if not adopted as law, these standards can often be used as evidence for standard of care.

Crew unity The concept that a fire company or unit shall remain together in a cohesive, identifiable working group to ensure personnel accountability and the safety of all members. A company officer or unit leader shall be responsible for the adequate supervision, control, communication, and safety of members of the company or unit.

Critical incident stress Stress associated with critical incidents, such as the injury or death of a coworker or a child.

Critical incident stress management (CISM) A process for managing the short- and long-term effects of critical incident stress reactions.

D

Demobilization The process of returning personnel, equipment, and apparatus after an emergency has been terminated.

F

Fire brigades The use of trained personnel within a business or at an industry site for firefighting and emergency response.

First in/last out The common approach often used at an emergency scene. Basically, the first arriving crews are generally the last to leave the scene.

Flashover A sudden, full involvement in flame of materials and gases within a room.

Flexibility The range of motion in a joint, which depends on the extensibility of soft tissues.

Freelancing Occurs when responders do not follow the incident plan at a scene and do what they want on their own. A failure to stay with assigned group.

Frequency How often a risk occurs or is expected to occur.

G

Goals Broad statements of what needs to be accomplished.

H

Haddon matrix A 4 × 3 matrix used to help analyze injuries and accidents in an attempt to determine processes designed to reduce them.

Heat transfer The transfer of heat through conduction, convection, radiation, and direct flame contact.

Hits Number of documents found when searching for a key word or phrase using a search engine.

I

Immediately dangerous to life and health (IDLH) Used by several OSHA regulations to describe a process or an event that could produce loss of life or serious injury if a responder is exposed to or operates in the environment.

Incident management system (IMS) An expandable management system for dealing with a myriad of incidents to provide the highest level of accountability and effectiveness. Limits span of control and provides a framework of breaking the big job down into manageable tasks.

Interior structural firefighting The physical activity of fire suppression, rescue, or both, inside buildings or enclosed structures that are involved in a fire beyond the incipient stage.

International Association of Fire Chiefs (IAFC) Organization of fire chiefs from the United States and Canada.

International Association of Fire Fighters (IAFF) Labor organization that represents the majority of organized firefighters in the United States and Canada.

Internet search engines Programs on the Internet that allow a user to search the entire Internet for key words or phrases.

L

Links Used on Internet pages so that each page may be directly tied to a page on another Internet site.

M

Matrix A chart used to categorize actions or events for analysis.

Muscular endurance The ability of the muscle to perform repeated contractions for a prolonged period of time; the ability of the muscle to persist.

Muscular strength The maximal amount of force a muscle or group of muscles can exert in a single contraction; the ability to apply force.

Mushrooming A situation in which heat and gases accumulate at the ceiling or top floor of a multistory building and then back down. Can be prevented with vertical ventilation.

N

National Association of Emergency Vehicle Technicians (NAEVT) An organization that bestows professional certification in many areas for persons involved in emergency service vehicle maintenance.

National Fire Incident Reporting System (NFIRS) The uniform fire incident reporting system for the United States; the data from this report is analyzed by the United States Fire Administration.

National Institute for Occupational Safety and Health (NIOSH) Federal agency responsible for conducting research and making recommendations for the prevention of work-related injury and illness; NIOSH is part of the Centers for Disease Control and Prevention (CDC).

NFPA 1403 The National Fire Protection Association's consensus standard on live fire training evolutions.

NFPA 1500 The National Fire Protection Association's consensus standard on fire department occupational safety and health.

O

Occupational disease An abnormal condition, other than an occupational injury, caused by an exposure to environmental factors associated with employment.

Occupational injury An injury that results from exposure to a single incident in the work environment.

Occupational Safety and Health Administration (OSHA) Federal agency tasked with the responsibility for the occupational safety of employees.

OSHA 300 Log of work-related injuries and illnesses.

OSHA 300-A Summary of work-related injuries and illnesses.

OSHA 301 The injury and illness incident report.

Outcome evaluation An evaluation that answers the question, "Did the program meet the expected goals?"

P

Peer defusing The concept of using a trained person from the same discipline to talk to an emergency responder after a critical incident serves as a means to allow the responder to talk about his or her feelings about the event in a nonthreatening environment.

Personal alert safety system (PASS) A device that produces a high-pitched audible alarm when the wearer becomes motionless for some period of time; useful for attracting rescuers to a downed firefighter.

Personnel accountability reports (PARs) A verbal or visual report to incident command or to the accountability officer regarding the status of operating crews. Should occur at specific time intervals or after certain tasks have been completed.

Postincident analysis (PIA) A critical review of the incident after it occurs. The postincident analysis should focus on improving operational effectiveness and safety.

Preventive maintenance (PM) program An ongoing program for maintenance on vehicles. Designed to provide routine care, oil changes and the like, as well as catch minor problems before they become major ones.

Process evaluation The ongoing evaluation of the various processes associated with a program or task.

R

Rapid interview companies (RIC) An assignment of a group of rescuers with the sole purpose of rapid deployment to reports of operating personnel in trouble or missing.

Regulations Rules or laws promulgated at the federal, state, or local level with a requirement to comply.

Rehabilitation The group of activities that ensures responders' health and safety at an incident scene. May include rest, medical surveillance, hydration, and nourishment.

Risk control A common approach to risk management in which measures and processes are implemented to help control the number and the severity of losses or consequences of risk to the organization.

Risk management A process of using the available resources of an organization to plan and direct the activities in the organization so that detrimental effects can be minimized.

Risk transfer The process of letting someone else assume the risk; for example, buying auto insurance transfers the consequences of an accident to the insurance company.

Risks The resultant outcome of exposure to a hazard.

Rollover The rolling of flame under the ceiling as a fire progresses to the flashover stage.

S

Search engines Programs on the Internet that allow a user to search the entire Internet for key words or phrases.

Self-contained breathing apparatus (SCBA) An atmosphere-supplying respirator for which the breathing air source is designed to be carried by the user.

Severity How severe the result is when a risk occurs.

Standard of care The concept of what a reasonable person with similar training and equipment would do in a similar situation.

Standard operating procedures (SOPs) Sometimes called standard operating guidelines, these are department-specific operational procedures, policies, and rules made to assist with standardized actions at various situations.

Standardized apparatus Apparatus that has exactly the same operation and layout of other similar apparatus in a department. For example, all of the department's pumpers would be laid out the same, operate the same, and have the same equipment. Useful for the situations when crews must use another crew's apparatus.

Standards Often developed through the consensus process, standards are not mandatory unless adopted by a governmental authority.

Stress The body's reaction to an event. Not all stress is bad; in fact, some level of stress is necessary to get a person to perform, for example, the stress associated with a report that is due is often the motivating factor in doing it.

T

Thermal Protective Performance rating (TPP) The amount of protection against both convective and radiant heat that your composite gear outer shell, thermal liner, and moisture barrier should be able to provide in the event of a flashover.

Tort Liability An action that harms another person or group and can occur when a person or group of people act, or fail to act, and thus harm another directly, or indirectly.

Total Heat Loss (THL) The measurement of an ensemble's heat stress reduction capability. The higher the THL value, the greater the benefit for the firefighter.

Transitional duty Sometimes called alternative duty or modified duty, these programs allow an injured employee to return to work, with restrictions, for some period of time while recuperating; this allows the organization to reduce lost time while utilizing the skills of the injured member.

U

Unified command Used in the IMS when two or more jurisdictions or agencies share incident command responsibilities but do so in conjunction with each other.

United States Fire Administration Department under the Federal Emergency Management Agency that directs and produces fire programs, research, and education.

Universal precautions Term used to describe the practice of treating all patients as if they carried an infectious disease, in terms of body fluid contamination.

V

Vehicle rescue Comprised of action to gain access, disentangle, and remove persons trapped in vehicles.

ACRONYMS

ADA	Americans with Disabilities Act		GIS	Geographic Information System
ADEA	Age Discrimination in Employment Act		GPS	Global positioning system
AFGP	Assistance to Firefighters Grant Program		HAZWOPER	Hazardous Waste Operations and Emergency Response
AHJ	Authorities having jurisdiction		HBV	Hepatitis B virus
ANSI	American National Standards Institute		HIPAA	Health Insurance Portability and Accountability Act
ASTM	American Society for Testing and Materials		HIV	Human immunodeficiency virus
			HMEP	Hazardous Materials Emergency Preparedness
BSI	Body substance isolation		HSO	Health and safety officer
CBRNE	Chemical, biological, radiological, nuclear, or explosive		HSSP	Homeland Security Standards Panel
CDC	Centers for Disease Control and Prevention		IAFC	International Association of Fire Chiefs
CFR	Code of Federal Regulations		IAFF	International Association of Fire Fighters
CISM	Critical incident stress management		IC	Incident commander
CVD	Cardiovascular disease		IDLH	Immediately dangerous to life and health
EMS	Emergency medical services		IMS	Incident management system
EPA	Environmental Protection Agency		ISO	Incident safety officer
			IV	Intravenous line
FFFIPP	Fire Fighter Fatality Investigation and Prevention Program		LEPC	Local emergency planning committee
FIRE	Fire Investment and Response Enhancement Act		NAEVT	National Association of Emergency Vehicle Technicians
FLSI	16 Firefighter Life Safety Initiatives			

NFIRS	National Fire Incident Reporting System
NFPA	National Fire Protection Association
NIOSH	National Institute for Occupational Safety and Health
NVFC	National Volunteer Fire Council
OSHA	Occupational Safety and Health Administration
OVC	Organic vapor canister
PAR	Personnel accountability reports
PASS	Personal alert safety system
PIA	Postincident analysis
PM	Preventive maintenance
PPE	Personal protective equipment
PSOB	Public Safety Officer's Benefit Program
RAID	Recognize, approach, identify, decide
RCFP	Rural Community Fire Protection
RIC	Rapid intervention company

RIT	Rapid intervention team
RPD	Recognition-primed decision
SAR	Supplied air respirator
SAFER	Staffing for Adequate Fire & Emergency Response Grants program
SCBA	Self-contained breathing apparatus
SOG	Standard operating guidelines
SOP	Standard operating procedure
THL	Total heat loss
TIP	Thermal imaging camera
TPP	Thermal protective performance rating
TWA	Time weighted average
USFA	United States Fire Administration
VCOS	Volunteer Chief Officer's Section
VFA	Volunteer Fire Assistance
WMD	Weapons of mass destruction

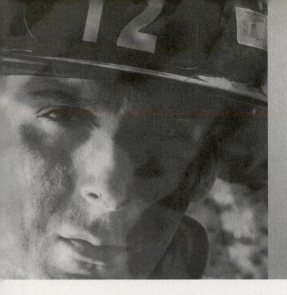

INDEX

A

accountability, on-scene, 129-131. *See also* roles and responsibilities

ACE (American Council on Exercise), 97

action plans, safety programs, 222-223

aerobic fitness, 99-100. *See also* fitness, physical

Age Discrimination in Employment Act (1967), 264

Agricolos, 3

AIDS, 51. *See also* bloodborne pathogens

aircraft fires, 115-117

airport fire fighters, 45

alcohol use and abuse, 101

ambulance safety, 79, 82-89

America Burning, 5-7

American Council on Exercise (ACE), 97

American National Standards Institute (ANSI), 49-50

American Society for Testing Materials, 50

Americans with Disabilities Act (1990), 263

animals, on-scene safety, 144

Annual Firefighter Death Report, 6

Annual Firefighter Injury Report, 6

ANSI (American National Standards Institute), 49-50

apparatus safety, overview, 79, 82-89

Assistance to Firefighters Grant Program, 266

ASTM International, 50

attitude, safety initiatives, 18

audits, safety, 58

automotive fire apparatus, 46-47, 47

B

back drafts, 115

behavioral fitness, 101

biological hazards
  interagency considerations, 102-104
  protective equipment, 47
  terrorism events, 169-170

bloodborne pathogens, 19, 35-36, 42, 47, 150-153

boats, swiftwater rescue, 56

body composition, 99-100

body substance isolation (BSI), 152-153

boots. *See* personal protective equipment (PPE)

budgets and funding, 264-266

C

carbon monoxide, 173

cardiovascular fitness, 99-100, 251

Case Study
  information management, 241-242
  live fire training, 24-25
  Mayday emergency, 195
  on-scene safety, 108, 141
  physical fitness, 260-261
  postincident safety, 179
  rapid intervention teams, 20

safety program evaluation, 233

single-family home fire, 2

standard operating procedures, 217-218

swiftwater rescue, 56

vehicle safety, 72-73

water rescue training death, 158

cause-and-effect diagram, 64

cave search and rescue, 164-165

Centers for Disease Control and Prevention (CDC), 42, 50-51, 257

chemical hazards
  interagency considerations, 102-104
  natural disasters, 172-173
  NFPA standards, 49
  terrorism events, 169-170

ChemTrec, 162

Child Protection Act (1966), 4

CISM (critical incident stress management), 101, 182, 189-191

civil disturbances, 168-169

Coal Mine Health and Safety Act (1969), 4

Coal Mine Safety Act (1952), 4

coat. *See* personal protective equipment (PPE)

Code of Federal Regulations, 26

codes, 27. *See also* National Fire Protection Association (NFPA)

cognitive, defined, 237